Wireless Networks

Series Editor
Xuemin Sherman Shen, University of Waterloo, Waterloo, Canada

The purpose of Springer's Wireless Networks book series is to establish the state of the art and set the course for future research and development in wireless communication networks. The scope of this series includes not only all aspects of wireless networks (including cellular networks, WiFi, sensor networks, and vehicular networks), but related areas such as cloud computing and big data. The series serves as a central source of references for wireless networks research and development. It aims to publish thorough and cohesive overviews on specific topics in wireless networks, as well as works that are larger in scope than survey articles and that contain more detailed background information. The series also provides coverage of advanced and timely topics worthy of monographs, contributed volumes, textbooks and handbooks.

Ruilong Deng · Zhenyong Zhang · Mengxiang Liu ·
Peng Cheng

Moving Target Defense in the Smart Grid

Springer

Ruilong Deng
Hangzhou, China

Zhenyong Zhang
Guiyang, China

Mengxiang Liu
Sheffield, UK

Peng Cheng
Hangzhou, China

ISSN 2366-1186 ISSN 2366-1445 (electronic)
Wireless Networks
ISBN 978-3-031-92377-7 ISBN 978-3-031-92378-4 (eBook)
https://doi.org/10.1007/978-3-031-92378-4

This Springer imprint is published by the registered company Springer Nature Switzerland AG
The registered company address is: Gewerbestrasse 11, 6330 Cham, Switzerland

Preface

The global pursuit of net-zero emissions and sustainable development, coupled with increasing energy demands and recurring energy crises, has accelerated the transformation of traditional power grids into smart grids. This evolution involves integrating advanced electronic devices and communication networks to enable environmentally friendly, real-time, and economically efficient control and operation. However, the widespread adoption of IP-based devices and communication networks exposes the smart grid to various cybersecurity vulnerabilities. Among these threats, false data injection (FDI) attacks pose significant risks to critical system operations, including state estimation, voltage control, and economic dispatch. This has prompted researchers to develop various defensive methods to enhance the cybersecurity of smart grids.

Traditional IT security approaches face significant limitations in protecting smart grids, particularly given the dynamic nature of adversarial environments characterized by zero-day vulnerabilities, insider threats, and supply chain risks. A key observation reveals that successful cyberattacks on smart grids heavily rely on attackers' knowledge of critical system parameters, such as grid topology and line configurations. This insight has led to the development of moving target defense (MTD), an innovative defensive mechanism that proactively perturbs these key parameters to prevent attackers from acquiring the information necessary for constructing effective attacks. As a significant complement to conventional IT security measures, MTD enhances smart grid cybersecurity by dynamically altering the attack surface, thereby increasing system uncertainty and complexity. This approach significantly reduces attackers' opportunities to conduct successful reconnaissance and launch undetected cyberattacks.

This monograph presents a comprehensive analysis of MTD, covering its performance evaluation, deployment optimization, and impact on smart grid stability. We begin with an extensive literature review that examines cyber MTD, physical MTD, and cyber-physical coordinated MTD approaches. Following this, we provide a thorough theoretical analysis validating MTD's effectiveness and completeness in detecting FDI attacks. We then analyze MTD's hiddenness from an attacker's perspective and propose a coordinated defense framework to enhance this aspect. To address the complexities arising from nonlinear AC state estimation, we develop sensitivity-based approximation methods to quantify MTD's effectiveness and hiddenness in AC power systems, and incorporate these metrics into an optimization framework that maximizes effectiveness while maintaining hiddenness. Furthermore, we conduct an in-depth analysis of MTD's effects on small signal stability, aiming to mitigate potential negative impacts while preserving defensive capabilities.

This monograph concludes by exploring future research directions and emphasizing the crucial role of MTD in achieving net-zero goals. We highlight how cost-effective MTD strategies can contribute to both grid security and sustainability objectives, as secure and reliable grid operation is fundamental to the integration of renewable energy sources and the realization of a low-carbon future. The research findings and design principles presented in this monograph provide valuable insights for practical MTD deployment and aim to inspire innovative approaches to smart grid defense strategies, ultimately supporting the transition toward a more sustainable and secure future energy infrastructure.

Hangzhou, China — Ruilong Deng
Guiyang, China — Zhenyong Zhang
Sheffield, UK — Mengxiang Liu
Hangzhou, China — Peng Cheng

Competing Interests The authors have no competing interests to declare that are relevant to the content of this manuscript.

Contents

Acronyms

AC	Alternating Current
AGC	Automatic Generation Control
AMI	Advanced Metering Infrastructure
AVR	Automatic Voltage Regulation
BDD	Bad Data Detection
CCPA	Coordinated Cyber-Physical Attack
CSD	Current Sharing Deviation
DC	Direct Current
DER	Distributed Energy Resources
D-FACTS	Distributed Flexible AC Transmission System
DFS	Depth-First Search
DMTDR	Decentralized Moving Target Defense via Data Replication
EMS	Energy Management System
EMTD	Extended Moving Target Defense
ENISA	European Union Agency for Cybersecurity
EXR-MTD	Explicit Residual-Based Moving Target Defense
FACTS	Flexible AC Transmission System
FDI	False Data Injection
FDIA	False Data Injection Attack
FRVM	Flexible Random Virtual IP Multiplexing
IED	Intelligent Electronic Device
IoT	Internet of Things
IP	Internet Protocol
IT	Information Technology
LRA	Load Redistribution Attack
MB-PSS	Multi-Band Power System Stabilizer
MPS	Measurement Protection Strategy
MT6D	Moving Target IPv6 Defense
MTD	Moving Target Defense
MTDper	Moving Target Defense-Induced Perturbation
NDRC	National Development and Reform Commission

NERC	North American Electrical Reliability Council
NESCOR	National Electric Sector Cybersecurity Organization Resource
NIST	National Institute of Standards and Technology
NWLS	Nonlinear Weighted Least Squares
OPF	Optimal Power Flow
PDDL	Proactive Distributed Detection and Localization
PFI	Power Flow Invariant
PFV	Physical Function Virtualization
PMU	Phasor Measurement Unit
PoP	Polynomial Optimization Problem
PQ	Active (P) and Reactive (Q) Power
PSS	Power System Stabilizer
PV	Active Power (P) and Voltage (V)
RAE	Relative Approximation Error
RTU	Remote Terminal Unit
SCADA	Supervisory Control and Data Acquisition
SDN	Software Defined Network
SE	State Estimation
VBD	Voltage Balancing Deviation
WAMPAC	Wide Area Monitoring, Protection, and Control

Chapter 1
Introduction

1.1 Background

1.1.1 The Smart Grid and Cybersecurity

As the world moves towards net-zero emissions, power systems are experiencing unprecedented transformations. The dual pressures of an intensifying energy crisis and rising electricity demand are catalyzing a shift from traditional mechanical operations to intelligent automation, fundamentally reshaping power grids into smart grids. These next-generation systems leverage cutting-edge communication [1, 2], computation, and control technologies to enable comprehensive monitoring, seamless interaction, and precise regulation across all power system components, ensuring efficient and reliable operation. However, the deep integration of IT and physical infrastructure that enables these capabilities also introduces new vulnerabilities. As shown in Fig. 1.1, hardware, software, communication channels, and human operators all present potential attack surfaces that malicious actors can exploit. Given its critical role in enabling smart grid functionality [3], the associated cybersecurity challenges have become a primary focus for both industry practitioners and academic researchers.

Unlike traditional IT systems [4] where cyberattacks primarily affect digital assets, attacks on smart grids can inflict catastrophic damage to physical infrastructure, potentially resulting in widespread power outages, severe economic losses, and threats to human safety. While traditional power grids focus mainly on functional safety through well-established protective measures, smart grids face unpredictable and difficult-to-control cybersecurity challenges. As illustrated in Fig. 1.2, cyberattacks on power systems have increased dramatically worldwide from 2010 to 2022, with a notable surge in the past three years. The power grid's status as critical national infrastructure has made it an attractive target for both state-sponsored actors and profit-motivated cybercriminals, as evidenced by the watershed 2015 Ukraine Blackout incident [6, 7]. More recent attacks, including the REvil [8] and

R. Deng et al., *Moving Target Defense in the Smart Grid*, Wireless Networks,
https://doi.org/10.1007/978-3-031-92378-4_1

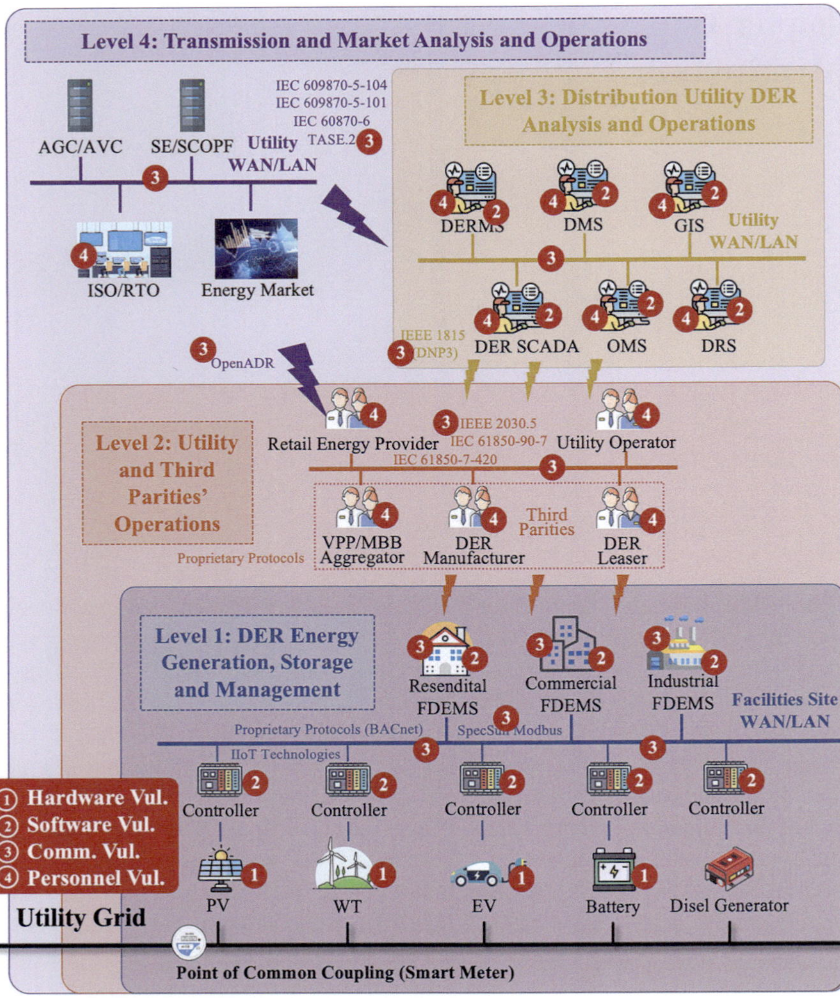

Fig. 1.1 Hierarchical framework of the smart grid as well as potential hardware, software, communication, and personnel vulnerabilities [5]

EnerCon [9] incidents, demonstrate a concerning trend of adversaries specifically targeting renewable energy infrastructure to extort ransom payments or sabotage vital communication networks. These escalating threats underscore the urgent need to protect smart grids against an expanding array of sophisticated cyberattacks.

Recognizing the critical importance of smart grid security, governments worldwide have established comprehensive regulatory frameworks to combat cybersecurity threats. NIST took a pioneering step in 2014 by introducing its "Guidelines for Smart Grid Cybersecurity", which presented an innovative layered defense strategy along with detailed methodologies for risk assessment and threat

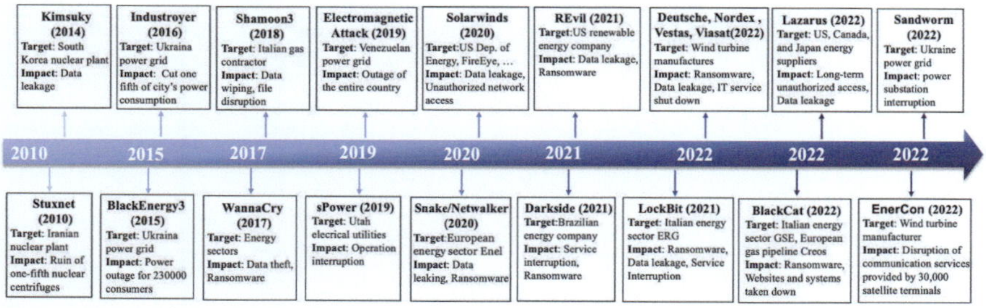

Fig. 1.2 Timeline of the cyberattacks targeting at the smart grid from 2010 to 2022 with an emphasis on the recent three years

quantification [10]. In parallel, ENISA released its "Smart Grid Security Recommendations" in 2013, emphasizing cyber-physical security and outlining robust protection measures including end-to-end authorization protocols and comprehensive supervision frameworks [11]. China has also made significant strides, with the NDRC introducing the concise yet powerful "secure zoning, dedicated networks, horizontal isolation, vertical authentication" principle in its 2014 "Regulations on Power Monitoring System Security Protection" [12]. This commitment to grid security was further reinforced in China's 13th Five-Year Plan (2016), which prioritized enhancing threat detection and continuous defense capabilities for critical infrastructure, particularly smart grids [13].

1.1.2 Vulnerability of State Estimation

The smart grid stands as a cornerstone of modern infrastructure and represents one of humanity's most sophisticated intelligent systems. At its core lies a bidirectional communication and centralized control architecture designed to ensure reliable and efficient operation. This bidirectional framework facilitates seamless data exchange across hierarchical levels, geographical regions, and diverse user groups, dramatically enhancing the grid's operational efficiency. The EMS serves as the central nervous system, enabling the control center to monitor the physical power system's status and orchestrate the complex interplay of generation, transmission, distribution, and consumption processes. Within the EMS, SE plays a pivotal role by augmenting the capabilities of the SCADA system. By processing raw SCADA data through transmission network models, SE generates optimal estimates of system state variables, providing a robust foundation for critical EMS applications including contingency analysis, AGC, load forecasting, and OPF [14].

Maintaining normal grid operations demands continuous real-time monitoring to precisely identify the system's operational state and enable swift responses to unexpected faults and incidents. The SCADA system functions as the EMS's sensory network, gathering comprehensive operational data through an array of field devices

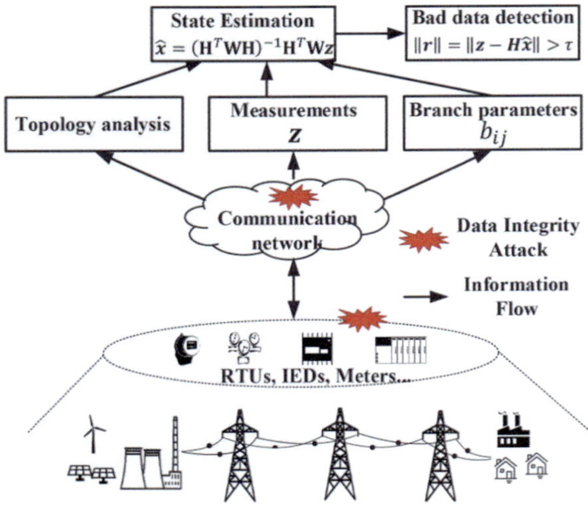

Fig. 1.3 Vulnerability of the communication structure of SE. The attacker can remotely access RTUs, IEDs, smart meters, and the communication infrastructure by exploiting their vulnerabilities, meanwhile, corrupt the measurements sent to the control center for SE

including RTUs, PMUs, and IEDs. This data encompasses critical parameters such as circuit breaker positions, relay states, transformer tap settings, and measurements of voltage, current, power flows, frequency, loads, and generator outputs. However, the raw data faces several inherent challenges: measurement errors, communication disruptions, and environmental interference can compromise accuracy; some power flows remain unmeasurable; and data transmission to the control center may be incomplete. To overcome these limitations, SE employs sophisticated algorithms to filter measurement uncertainties and construct a reliable, accurate representation of the grid's operational state (Fig. 1.3).

False Data Injection Attack

FDI attacks represent a sophisticated class of cyberattacks targeting the SE process in smart grids. First introduced by Liu et al. [15] at the 2009 ACM Conference on Computer and Communications Security (CCS), these attacks have since become the subject of extensive research examining their design under various conditions. The fundamental mechanism involves manipulating measurement values within the SCADA system to inject biases into state estimation results. These manipulated values cause the calculated system state to deviate from reality, potentially leading operators to make incorrect operational decisions. Based on the attacker's level of system knowledge, design methodology, and analytical approach, existing research on FDI attack can be categorized into four main areas:

(1) Sparse FDI Attack: When attackers possess complete knowledge of system topology and transmission line parameters, Liu et al. [15] established a theoretical framework for designing stealthy FDI attacks using the relationship $a = Hc$, where a represents the attack vector, H denotes the measurement matrix, and c indicates

the injected state bias. Building on these findings, researchers have explored methods for designing sparse FDI attacks that minimize attack resources while accounting for system protection mechanisms [16–23].

(2) Data Framing FDI Attack: This novel variant of FDI attacks exploits the BDD mechanisms within SE. While these attacks can be detected by the BDD process, they still compromise SE accuracy by manipulating the BDD to incorrectly identify legitimate measurements as anomalous data. When these valid measurements are removed, the SE calculations proceed using the compromised data, preventing operators from accurately assessing the system's true state [24].

(3) Partial-Information FDI Attack: Given the complexity of large-scale power systems, attackers rarely obtain complete system information such as circuit breaker positions, transformer settings, or comprehensive topology and transmission line parameters. Consequently, researchers have investigated more realistic scenarios where attackers design FDI attacks using only partial system information [25–32].

(4) Data-Driven FDI Attack: Recognizing that transmission network topology frequently adjusts based on operational conditions, and line parameters vary due to factors like OPF, AGC, and environmental conditions, some researchers have focused on designing FDI attacks using intercepted measurement data directly. This approach circumvents the need for precise physical parameter knowledge [33–38].

1.2 Moving Target Defense

The MTD is a new type of defense strategy. Its core idea is to increase the cost and complexity of attacks by creating diverse mechanisms, deploying heterogeneous systems, and dynamically changing parameters, thereby reducing the resources that attackers can exploit and lowering the probability of successful attacks, thus enhancing the system's robustness. Since the smart grid integrates information technology with physical systems, network attacks from information systems can cause damage to the physical systems of the smart grid or affect economic behaviors in the electricity market. Currently, extensive research and security events show that smart grids face a variety of security threats. Among them, as described in the previous subsection, FDI attacks are a highly destructive form of attack. Attackers can use acquired information about the grid topology and transmission line parameters to design covert attack vectors targeting SE methods and BDD, altering SE results, and misleading system operators' decisions, analyses, and controls.

Investigation shows that for achieving covert FDI attacks, it is assumed that the attacker knows the system topology and transmission line parameters. Therefore, some researchers [39, 40] suggest that by actively disturbing the transmission line parameters, attackers can be prevented from obtaining the necessary system data, making it difficult for them to design FDI attacks. This strategy relies on deploying D-FACTS on transmission lines. By configuring data within the D-FACTS devices, transmission line parameters can be dynamically changed. Since attackers cannot predict the changes in transmission line parameters, this increases the uncertainty in the information available for designing FDI attacks.

Divan and Johal [41] were the first to introduce D-FACTS devices into power systems to control power flow on transmission lines. Operators can use D-FACTS devices to alter the impedance of transmission lines, thereby controlling the power flow capacity distribution and eliminating certain transmission line capacity constraints and bottlenecks in the transmission network. D-FACTS devices are small and lightweight, capable of being connected in series on transmission lines, and their excellent communication, control, and computing capabilities make them highly promising for application [42].

In recent years, power system security experts have incorporated D-FACTS into power system SE to defend against covert FDI attacks [39, 40, 43–48], and this strategy of disturbing transmission line parameters to achieve defensive effects has been named MTD, with its defense and implementation framework shown in Fig. 1.4.

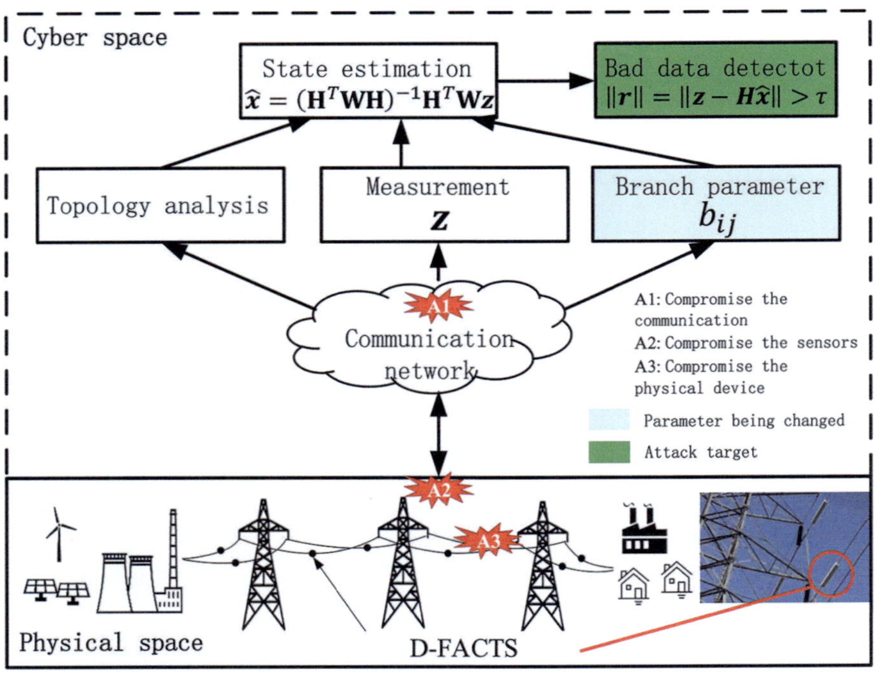

Fig. 1.4 The MTD framework and its implementation based on D-FACTS devices

1.2.1 Comparison Between Moving Target Defense and Traditional Defense Strategies

Traditional defense strategies for smart grids typically address cyber and physical security independently. This siloed approach, while established, fails to account for the deeply interconnected nature of modern power systems. Here, we analyze the limitations of traditional approaches and highlight how MTD offers a more comprehensive defense framework.

(1) Cyber-side Defense: Smart grid networks face unique security challenges that make traditional IT security measures insufficient. Unlike personal computers that can readily employ real-time antivirus protection, critical grid components often cannot accommodate security software due to operational constraints—running virus scans or updates could trigger equipment downtime or system failures.

While the smart grid employs fundamental security measures like network segmentation, dedicated networks, horizontal isolation, and vertical authentication, these approaches face significant challenges:

- Key management becomes increasingly complex due to the vast number of heterogeneous devices (smart meters, RTUs, and IEDs) requiring secure authentication.
- Traditional security tools (blacklists, firewalls, and intrusion detection) remain vulnerable to zero-day exploits.
- Device spoofing can bypass even sophisticated authentication mechanisms.

Recent research has explored incorporating physical-layer security features, such as:

- Deep packet inspection leveraging physical semantics in function codes [49].
- Machine learning approaches that analyze SCADA data for anomaly detection.
- Physics-based rules for validating state transitions [50].

However, these methods still face limitations: machine learning approaches depend heavily on clean training data and carefully selected features, while physics-based validation requires precise system modeling.

(2) Physical-side Defense: Traditional power grid protection focuses on maintaining functional safety through mechanisms like:

- Relay protection for transmission line faults.
- Circuit breaker isolation of compromised components.
- Remote and bus protection schemes.

While effective for containing physical faults, these localized protection measures cannot adequately address coordinated cyber attacks. Moreover, protection systems themselves can become attack surfaces if compromised. System-theoretic approaches analyzing stability and robustness offer additional protection layers. However, their abstract models often fail to capture real operational patterns, leaving vulnerabilities. Even redundancy mechanisms, while providing fault tolerance, can be exploited to construct more sophisticated attacks.

(3) Cyber-Physical Collaborative Defense: MTD represents a paradigm shift by integrating cyber and physical defense mechanisms. Unlike traditional approaches, MTD:

- Leverages physical perturbations as probes to detect cyber attacks.
- Creates dynamic uncertainty that increases attack complexity.
- Exploits attackers' inability to precisely track physical parameter changes.
- Provides active rather than passive defense.

By injecting controlled physical uncertainty while monitoring cyber-side responses, MTD can detect attacks that might evade traditional defenses. This cyber-physical integration is particularly effective for protecting SE, as attackers struggle to maintain accurate system models in the face of dynamic physical changes.

The effectiveness of MTD stems from its alignment with the fundamental nature of smart grids as cyber-physical systems. Rather than treating cyber and physical security as separate domains, MTD creates a unified defense framework that exploits the interdependencies between these layers. This holistic approach represents a significant advancement over traditional security measures in addressing modern smart grid threats.

1.2.2 Research Challenges

Despite extensive research on smart grid SE security, including the perspectives from both the attacker and defender, significant limitations still remain as discussed in the following.

Limitations from the Attacker Perspective

Current research makes overly strong assumptions about attacker capabilities, primarily focusing on two attack approaches:

- **Model-based attacks** require detailed knowledge of system topology and transmission line parameters. This assumption is unrealistic as these parameters frequently change due to system adjustments and environmental factors, making attack design challenging.
- **Data-driven attacks** depend on access to substantial historical data [54]. However, incomplete data and limitations in feature selection often render such attacks ineffective.

Challenges from the Defender Perspective

From the defender's viewpoint, four key limitations exist:

- **Vulnerability of static defenses**: Traditional measures like encryption and secure PMUs cannot guarantee long-term security. Given sufficient time, attackers can develop strategies to circumvent these protections. Model-based detection mechanisms are particularly vulnerable due to their inability to accurately characterize dynamic system features.
- **Limited adaptability to system changes**: Detection methods that rely on stable system operational states struggle when conditions change. Statistical approaches require consistent operational states for accurate feature extraction, while machine learning methods need extensive training across various scenarios, complicating real-time defense implementation.
- **Lack of cyber-physical integration**: Most current defenses focus solely on information-side analysis, overlooking potential physical-side defensive capabilities. Smart grids are cyber-physical systems and integrating both domains could enable more robust defense strategies.
- **Unresolved MTD optimization challenges**:
  - Quantifying MTD effectiveness and hiddenness.
  - Optimizing D-FACTS device deployment for cost efficiency.
  - Investigating the impact of MTD on power system performance.

1.3 Aim of the Monograph

This monograph aims to design practical MTD strategies for both AC and DC power systems with strong effectiveness and hiddenness. Moreover, the impacts of MTD perturbation on operation cost and control performance are systematically investigated.

Firstly, in Chap. 2, we present a comprehensive review to elaborate existing MTD strategies in cyber-physical power systems. In particular, the cyber MTD strategies such as IP/port address shuffling and network topology mutation are carefully surveyed. Utilizing the flexibility of modern devices and programmable controllers, the physical MTD strategies have been broadly applied to power grid SE and microgrid control. Finally, a representative cyber-physical coordinated MTD strategy that creates physical function virtualization is illustrated to showcase the integration of both cyber and physical technologies.

Secondly, in Chap. 3, we prove that, an MTD has the capability to thwart all FDI attacks constructed with former branch susceptances only if (i) the number of branches l in the power system is not less than twice that of the system states n (i.e., $l \geq 2n$, where $n + 1$ is the number of buses); (ii) the susceptances of more than n branches, which cover all buses, are perturbed [51]. Moreover, we prove that the state variable of a bus that is only connected by a single branch (no matter it is perturbed or not) can always be modified by the attacker. Nevertheless, in order to reduce the attack opportunities of potential attackers, we first exploit the impact of the susceptance perturbation magnitude on the dimension of the *stealthy attack space*, in which the attack vector is constructed with former branch susceptances. Then, we propose that, by perturbing an appropriate set of branches, we can minimize the dimension of the *stealthy attack space* and maximize the number of covered buses. Besides, we consider the increasing operation cost caused by the activation of MTD. Finally, we conduct extensive simulations to validate our findings with IEEE standard test power systems.

Thirdly, in Chap. 4, to uncover the insight of MTD's hiddenness, we study the conditions needed for achieving a hidable MTD. We find that the hiddenness of MTD is closely related to the branch perturbations, system topology, and attacker's knowledge. From the attacker's perspective, we prove that an MTD can be detected by the attacker only if he/she knows the previous parameters of a set of branches that forms a circle and the measurements corresponding to those branches after MTD. But once the attacker has full knowledge of branch parameters before MTD and has obtained all measurements after MTD, it is proved that we can never achieve a hidable and effective MTD. From the defender's perspective, since it is impossible to know the attacker's capability, we cannot determine whether a constructed MTD is hidable or not by purely depending on the MTD design. To address this issue, we propose that, by protecting a basic set of measurements, we still can achieve a hidable and effective MTD regardless of the changes of power flows, the attacker's knowledge, and the branch perturbations. Furthermore, we validate our findings with the IEEE standard test power systems.

Fourthly, in Chap. 5, we attempt to conduct an explicit analysis on the MTD performance in AC power systems. First, we derive explicit approximations of measurement residuals to quantify the two metrics. Then, based on the projection matrix, maximizing the effectiveness is transformed to maximizing the lower bound of the approximated residual, under which the matrix inverse issue is addressed. Moreover, the maximization of hiddenness is achieved by the minimization of the approximated power flow difference caused by reactance perturbation. To balance the trade-off between effectiveness and hiddenness, the design of EXR-MTD is accomplished by aggregating the two sub-problems with an appropriate weight. Finally, extensive simulations are conducted to validate the performance of EXR-MTD. Numerical results indicate that EXR-MTD performs better than existing MTD strategies in terms of hiddenness, while the effectiveness of EXR-MTD is comparable to those of existing MTD strategies.

Fifthly, in Chap. 6, we deeply analyze the effect of MTD on the small signal stability and aim to alleviate the negative impact and guarantee its defending performance simultaneously. First, the stability is formally described using the eigenvalue sensitivity. The relationship between the MTD-induced perturbation (MTDper) and the stability criteria is derived. Second, a new indicator is proposed to measure the effectiveness of MTDper. Third, a constrained optimization problem is formulated to compute the bound of MTDper for guaranteeing the small signal stability. In addition, a surprising finding is that the stability margin can be improved and enhanced by optimizing the value of MTDper without losing the MTD's effectiveness. Finally, we evaluate the performance of MTDper and its impact on the small signal stability with extensive simulations on the IEEE 30-bus, 39-bus, and 68-bus test power systems.

Finally, we summarize this monograph and make future perspectives for the design of cost-efficient MTD strategies that are able to ensure the security and resilience of future net-zero energy transition.

References

1. X. Shen, J. Gao, W. Wu, M. Li, C. Zhou, W. Zhuang, Holistic network virtualization and pervasive network intelligence for 6G. IEEE Commun. Surv. Tutor. **24**(1), 1–30 (2022). 1st Quarter. https://doi.org/10.1109/COMST.2021.3135829
2. W. Wu, M. Li, K. Qu, C. Zhou, X. Shen, W. Zhuang, X. Li, W. Shi, Split learning over wireless networks: parallel design and resource management. IEEE J. Sel. Areas Commun. **41**(4), 1051–1066 (2023). https://doi.org/10.1109/JSAC.2023.3242704. (Apr.)
3. W. Wu, C. Zhou, M. Li, H. Wu, H. Zhou, N. Zhang, X. Shen, W. Zhuang, AI-native network slicing for 6G networks. IEEE Wirel. Commun. **29**(1), 96–103 (2022). https://doi.org/10.1109/MWC.001.2100338. (Feb.)
4. X. Shen, J. Gao, W. Wu, K. Lyu, M. Li, W. Zhuang, X. Li, J. Rao, AI-assisted network-slicing based next-generation wireless networks. IEEE Open J. Veh. Technol. **1**(1), 45–66 (2020). https://doi.org/10.1109/OJVT.2020.2965100. (Jan.)
5. M. Liu, F. Teng, Z. Zhang, P. Ge, M. Sun, R. Deng, P. Cheng, J. Chen, Enhancing cyber-resiliency of DER-based smart grid: a survey. IEEE Trans. Smart Grid **15**(5), 4998–5030 (2024)
6. E-ISAC, Analysis of the cyber attack on the Ukrainian power grid. SANS Industrial Control Systems Security Blog, Washington, DC, USA, Tech. Rep., 2016. [Online]. http://ics.sans.org/media/E-ISAC_SANS_Ukraine_DUC_5.pdf
7. G. Liang, S.R. Weller, J. Zhao, F. Luo, Z.Y. Dong, The 2015 Ukraine blackout: implications for false data injection attacks. IEEE Trans. Power Syst. **32**(4), 3317–3318 (2017)
8. B. Eckhouse, K. Mehrotra, Clean-energy giant invenergy suffers hack claimed by revil (2023). [Online]
9. Modern Power Systems, Enercon turbines disrupted by satellite cyber attack (2024). [Online]
10. The Smart Grid Interoperability Panel-Smart Grid Cybersecurity Committee, *Guidelines for Smart Grid Cybersecurity [R]* (National Institute of Standards and Technology, 2014)
11. E. Egozcue, D. Herreras Rodriguez, J. Alonso Ortiz, V. Fidalgo Villar, L. Tarrafeta, *Smart Grid Security [R]* (European Network and Information Security Agency, 2012)
12. National Development and Reform Commission (NDRC), *Implementation Measures for the Electricity Regulatory Ordinance [EB/OL]* (Decree of the People's Republic of China National Development and Reform Commission, 2014)
13. National Development and Reform Commission (NDRC), *Outline of the 13th Five-Year Plan [R]*, in *The 4th Session of the 12th National People's Congress* (2016)
14. Ali Abur and Antonio Gomez Exposito, *Power System State Estimation: Theory and Implementation [M]* (Marcel Dekker Inc., New York, 2004)
15. Y. Liu, P. Ning, M.K. Reiter, False data injection attacks against state estimation in electric power grids. ACM Trans. Inf. Syst. Secur. (TISSEC) **14**(1), 1–33 (2011)
16. O. Kosut, L. Jia, R.J. Thomas, L. Tong, On malicious data attacks on power system state estimation, in *IEEE UPEC*, pp. 1–6 (2010)
17. O. Kosut, L. Jia, R.J. Thomas, L. Tong, Malicious data attacks on smart grid state estimation: attack strategies and countermeasures, in *IEEE SmartGridComm*, pp. 220–225 (2010)

18. O. Kosut, L. Jia, R.J. Thomas, L. Tong, Limiting false data attacks on power system state estimation, in *IEEE CISS*, pp. 1–6 (2010)
19. G. Dan, H. Sandberg, Stealth attacks and protection schemes for state estimators in power systems, in *IEEE SmartGridComm*, pp. 214–219 (2010)
20. A. Giani, E. Bitar, M. Garcia, M. McQueen, P. Khargonekar, K. Poolla, Smart grid data integrity attacks: characterization and countermeasures, in *IEEE SmartGridComm* (2011)
21. M. Ozay, I. Esnaola, F.T.Y. Vural, S.R. Kulkarni, H.V. Poor, Sparse attack construction and state estimation in the smart grid: centralized and distributed models. IEEE J. Sel. Areas Commun. 1306–1318 (2013)
22. Q. Yang, J. Yang, W. Yu, D. An, N. Zhang, W. Zhao, On false data-injection attacks against power system state estimation: modeling and countermeasures. IEEE Trans. Parallel Distrib. Syst. 717–729 (2014)
23. J. Hao, R.J. Piechocki, D. Kaleshi, W.H. Chin, Z. Fan, Sparse malicious false data injection attacks and defense mechanisms in smart grids. IEEE Trans. Ind. Inform. 1198–1209 (2015)
24. J. Kim, L. Tong, R.J. Thomas, Data framing attack on state estimation. IEEE J. Sel. Areas Commun. 1460–1470 (2014)
25. M.A. Rahman, H. Mohsenian-Rad, False data injection attacks with incomplete information against smart power grids, in *IEEE GLOBECOM*, pp. 3153–3158 (2012)
26. X. Liu, Z. Li, Local load redistribution attacks in power systems with incomplete network information. IEEE Trans. Smart Grid 1665–1676 (2014)
27. Y. Sun, W.-T. Li, W. Song, C. Yuen, False data injection attacks with local topology information against linear state estimation, in *IEEE ISGT ASIA*, pp. 1–5 (2015)
28. X. Liu, Z. Bao, D. Lu, Z. Li, Modeling of local false data injection attacks with reduced network information. IEEE Trans. Smart Grid 1686–1696 (2015)
29. X. Liu, Z. Li, Local topology attacks in smart grids. IEEE Trans. Smart Grid **8**(6), 2617–2626 (2017)
30. R. Deng, H. Liang, False data injection attacks with limited susceptance information and new countermeasures in smart grid. IEEE Trans. Ind. Inform. 1619–1628 (2019)
31. J. Zhang, Z. Chu, L. Sankar, O. Kosut, Can attackers with limited information exploit historical data to mount successful false data injection attacks on power systems? IEEE Trans. Power Syst. 4775–4786 (2018)
32. K. Pan, A. Teixeira, M. Cvetkovic, P. Palensky, Cyber risk analysis of combined data attacks against power system state estimation. IEEE Trans. Smart Grid 3044–3056 (2019)
33. Y. Huang, M. Esmalifalak, H. Li, L. Song et al., Bad data injection in smart grid: attack and defense mechanisms. Cyber Secur. Smart Grid Commun. Part 2 27–33 (2013)
34. M. Esmalifalak, H. Nguyen, R. Zheng, Z. Han, Stealth false data injection using independent component analysis in smart grid, in *IEEE SmartGridComm*, pp. 244–248 (2011)
35. Z.-H. Yu, W.-L. Chin, Blind false data injection attack using PCA approximation method in smart grid. IEEE Trans. Smart Grid 1219–1226 (2015)
36. J. Kim, L. Tong, R.J. Thomas, Subspace methods for data attack on state estimation: a data driven approach. IEEE Trans. Signal Process. 1102–1114 (2015)
37. I. Markwood, Y. Liu, K. Kwiat, C. Kamhoua, Electric grid power flow model camouflage against topology leaking attacks, in *IEEE INFOCOM*, pp. 1–9 (2017)
38. W.-L. Chin, C.-H. Lee, T. Jiang, Blind false data attacks against AC state estimation based on geometric approach in smart grid communications. IEEE Trans. Smart Grid 6298–6306 (2018)
39. K.L. Morrow, E. Heine, K.M. Rogers, R.B. Bobba, T.J. Overbye, Topology perturbation for detecting malicious data injection, in *IEEE Hawaii ICSS*, pp. 2104–2113 (2012)
40. K.R. Davis, K.L. Morrow, R. Bobba, E. Heine, Power flow cyber attacks and perturbation-based defense, in *IEEE SmartGridComm*, pp. 342–347 (2012)
41. D. Divan, H. Johal, Distributed FACTs: a new concept for realizing grid power flow control. IEEE Trans. Power Electron. 2253–2260 (2007)
42. H. Johal, D. Divan, Design considerations for series-connected distributed FACTS converters. IEEE Trans. Ind. Appl. 1609–1618 (2018)

43. M.A. Rahman, E. Al-Shaer, R.B. Bobba, Moving target defense for hardening the security of the power system state estimation, in *ACM MTD*, pp. 59–68 (2014)
44. J. Tian, R. Tan, X. Guan, T. Liu, Enhanced hidden moving target defense in smart grids. IEEE Trans. Smart Grid 2208–2223 (2018)
45. S. Lakshminarayana, D.K. Yau, Cost-benefit analysis of moving-target defense in power grids, in *IEEE/IFIP DSN*, pp. 139–150 (2018)
46. K. Kuntz, M. Smith, K. Wedeward, M. Collins, Detecting, locating, & quantifying false data injections utilizing grid topology through optimized D-FACTS device placement, in *IEEE NAPS*, pp. 1–6 (2014)
47. H. Salehghaffari, F. Khorrami, Resilient power grid state estimation under false data injection attacks, in *IEEE PES ISGT*, pp. 1–5 (2018)
48. B. Li, G. Xiao, R. Lu, R. Deng, H. Bao, On feasibility and limitations of detecting false data injection attack on power grid state estimation using D-FATCS devices. IEEE Trans. Ind. Inform. 854–864 (2020)
49. J. Meng, Z. Yang, Z. Zhang, Y. Geng, R. Deng, P. Cheng, J. Chen, J. Zhou, SePanner: analyzing semantics of controller variables in industrial control systems based on network traffic, in *ACM ACSAC*, pp. 310–323 (2023)
50. Z. Yang, L. He, P. Cheng, J. Chen, D.K.Y. Yau, L. Du, PLC-Sleuth: detecting and localizing PLC intrusions using control invariants, in *IEEE RAID*, pp. 333–348 (2020)
51. Z. Zhang, R. Deng, D.K.Y. Yau, P. Cheng, J. Chen, Analysis of moving target defense against false data injection attacks on power grid. IEEE Trans. Inf. Forensics Secur. 2320–2335 (2020)
52. Z. Zhang, R. Deng, D.K.Y. Yau, P. Cheng, J. Chen, On hiddenness of moving target defense against false data injection attacks on power grid. ACM Trans. Cyber Phys. Syst. 25–53 (2020)
53. M. Liu, C. Zhao, Z. Zhang, R. Deng, Explicit analysis on effectiveness and hiddenness of moving target defense in AC power systems. IEEE Trans. Power Syst. 4732–4746 (2022)
54. Z. Zhang et al., Vulnerability of machine learning approaches applied in IoT-based smart grid: a review. IEEE Internet Things J. **11**(11), 18951–18975 (2024). https://doi.org/10.1109/JIOT.2024.3349381

Chapter 2
Literature Review of Moving Target Defense in the Smart Grid

2.1 Cyber Moving Target Defense

The cyber MTD strategies aim to perturb the cyber components such as hosts, virtual machines, and applications to increase the complexity of cyber environment and decrease the attack's success probability. For example, the IP mutation, software migration, and platform switch can effectively change the attack surface, thus making the critical information collected during reconnaissance and scanning phase obsolete, which may thwart the further propagation from an office host to another critical upper host or impede the preparation of denial-of-service attacks. Numerous insightful surveys [7, 8] have been presented to summarize the MTD operational schemes and focus on *what and when to perturb as well as how to assess*. This chapter choose to pay special attention to the MTD strategies deriving from the communication network layer as they have much closer relations to the physical layer of smart grid, while the MTD strategies in a broad sense based on web application, software, hardware are out of this monograph's scope. In particular, the surveyed cyber MTD strategies are divided into: (1) IP/port address shuffling, and (2) network topology mutation.

2.1.1 IP/Port Address Shuffling

IP address shuffling is one of the most common MTD strategies that randomize the attack surface, increasing the complexity of launching successful scanning attack and worm propagation. Benefiting from the vastly larger address space of IPv6 compared with that of IPv4, an MT6D strategy was proposed to prevent the attacker from targeting specific addresses by dynamically rotating network and transport layer addresses without affecting preexisting connections [14]. Govindarasu et al. [17, 18] applied end-to-end IP-hopping technique to the SCADA communication network of smart grid to deter reconnaissance attacks. As shown in Fig. 2.1, the proposed

R. Deng et al., *Moving Target Defense in the Smart Grid*, Wireless Networks,
https://doi.org/10.1007/978-3-031-92378-4_2

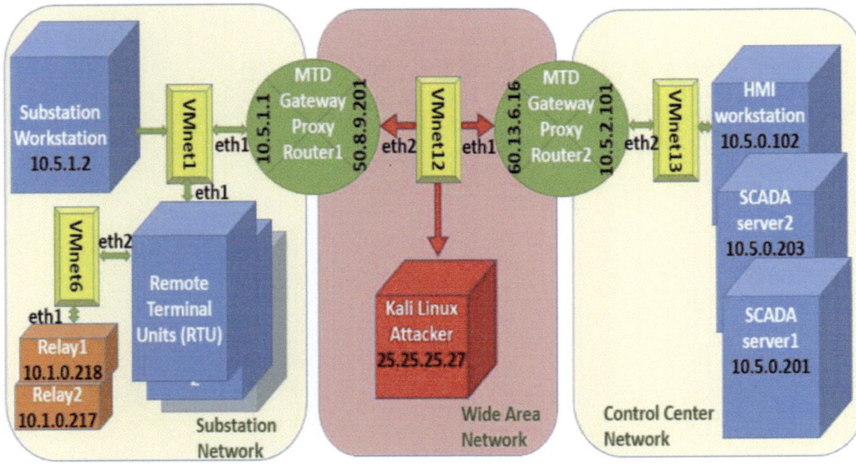

Fig. 2.1 The architecture of experimental setup for MTD IP hopping [17, 18]

architecture was implemented in Iowa State's PowerCyber testbed for targeted cyber-attacks, where the MTD routers work as dynamic proxies to mutate the IPs of external interfaces while still providing transparent, uninterrupted and seamless end-to-end SCADA communications. Massive studies were conducted to disclose the delay and throughput under different hopping rates in the realistic testbed, and recommendations were given to choose appropriate hopping rates to trade off the quality of service and security enhancement.

As demonstrated in Fig. 2.2, Sharma et al. [1] introduced an FRVM method based on SDN technology. The framework includes end-hosts, an SDN controller, SDN switches, as well as data and control planes. The active services running on a server host are identified by the port numbers, and a virtual IP is assigned to each service. The SDN controller handles all the multiplexing/de-multiplexing from virtual/real IP addresses to real/virtual IP addresses and the continuously updating flow table within SDN switches. The real IP addresses remain unchanged to avoid disruptions on user services, while the routing among virtual addresses is hidden and restricted in the programmable network, which can be thus kept transparent to the user.

In a nutshell, the FRVM enables a host machine to have multiple, random, time-varying virtual IP addresses, which are multiplexing to a real IP address of the host, eventually making the collected critical knowledge outdated and disturbing the scanning strategy.

MacFarland and Shue [2] applied an SDN-enabled IP mutation scheme to distinguish trustworthy clients from untrustworthy ones while circumventing the scalability issue for large-scale network. Antonatos et al. [3] devised a network harden strategy to counter against hitlist worms by randomizing the host IP addresses in a frequent and autonomous manner. The frequently updated IP addresses can render the hitlist information stale and force the worms to exhibit features that make them

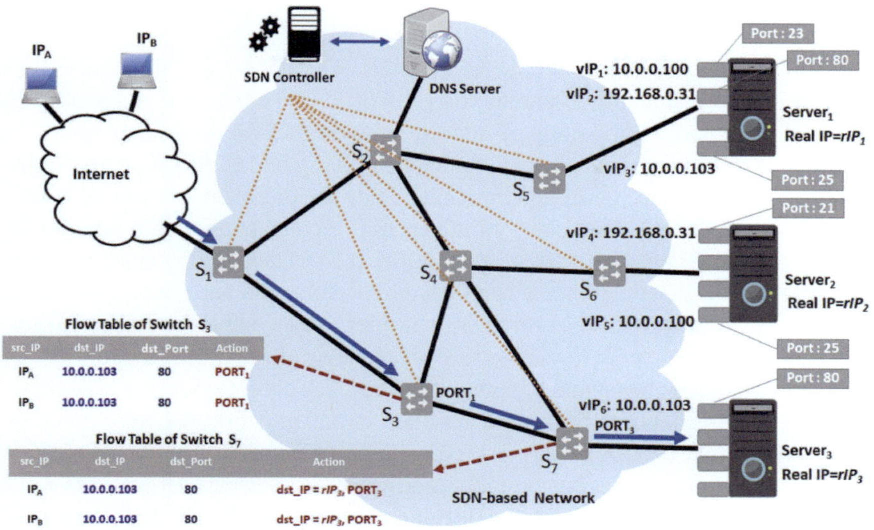

Fig. 2.2 The implementation of FRVM in SDN environment [1]

easier to contain at the perimeter. Jafarian et al. [4] presented a Openflow-enabled randomization approach to transparently mutate the virtual host IP addresses with high unpredictability and rate such that expected security enhancement against scanning attacks and worm propagation can be attained. The allocation of virtual IP addresses is formulated as a constraint satisfaction problem to maintain the configuration integrity and minimize incurred overhead.

Carroll et al. [5] introduced probabilistic models to quantify the attacker's success probability in terms of network size, quantity of scanned addresses, quantity of vulnerable systems, and the frequency of shuffling actions. Theoretical investigation was conducted to validate that the IP mutation is an acceptable defense approach when the number of compromised systems is limited. Kampanakis et al. [6] provided an unified MTD framework based on the programmable SDN technology, which mainly consist of a SDN controller that abstracts the network states, an adaption engine that injects random network adaption at regular intervals, and an analytic engine that assess the network's exposure to cyber attack events. Luo et al. [11] proposed to utilize port hopping, which constantly switches the service ports, as a MTD strategy to thwart reconnaissance attacks, vulnerability exploitation, and privilege escalation, while the impact of port hopping actions on quality of service has not been well studied [12].

2.1.2 Network Topology Mutation

Network topology mutation is another division of MTD strategies that dynamically change the networking topology of nodes to harden malicious eavesdropping and scanning attacks, as well as identify malicious probing nodes. Stefan et al. [9] presented a reconnaissance deception system to delay the discovery process of vulnerable hosts and identify the source of adversarial reconnaissance in a network by simulating virtual network topologies via utilizing SDN technology and honeypots. According to Fig. 2.3, the presented architecture of reconnaissance deception system consists of five components: (1) An SDN controller is responsible for dynamical generation and management of flow rules to steer and control and network traffic as well as identify malicious behaviors of network hosts, (2) A deception server is deployed to manipulate the network traffic and simulate virtual network resources considering specific policies, (3) A honeypot server simulates forged honeypot nodes, (4) A virtual network view generator provides a description of virtual network resources and their connectivity, and (5) Traffic queuing module controls link delay and bandwidth. When a data packet arrives at an SDN controller connected to the reconnaissance deception system, it deploys a flow rule aligning with the virtual network view description that either forwards the packet to the deception server, or sends the packet to its destination host after tags are added to the packet and addresses are translated. If a packet is forwarded to a real end host or honeypot, artificial delay based on appropriate traffic queuing disciplines is added to the replay packet to guarantee the

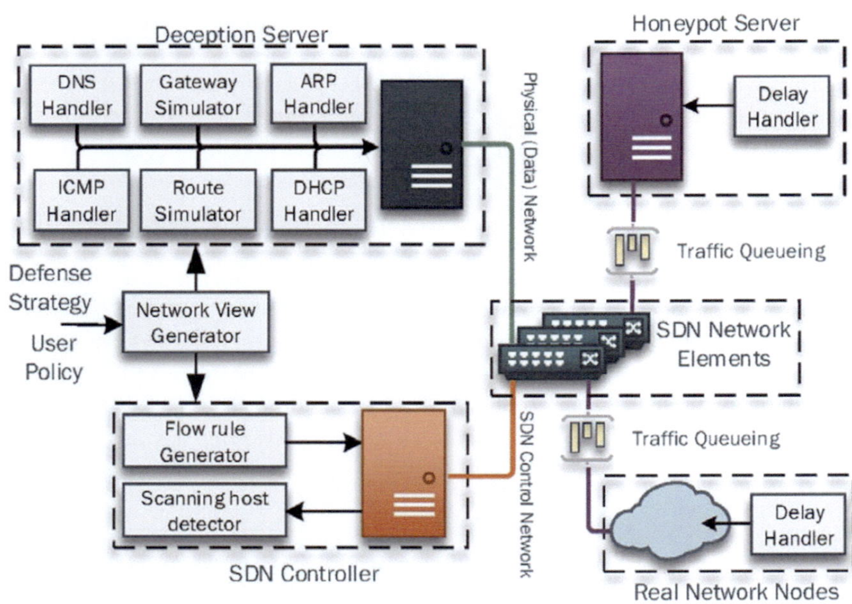

Fig. 2.3 The architecture of reconnaissance deception system [9]

consistency with other data packets. In addition to the SDN technology, Chin et al. advised the utilization of Docker container to mimic a virtual network hosts, which was validated to have satisfactory deterring and mitigation performance when in conjunction with appropriate intrusion detection strategies [15].

Hong et al. [10] formulated the optimal network topology reconfiguration as a shuffle assignment problem to maximize the network security against privilege escalation attacks, and a heuristic algorithm was proposed to resolve the scalability issue with negligible optimal gap. In particular, the shuffle assignment problem aims to maximize the number of variants to be compromised through reconfiguring the network topology. Exhaustive search is able to attain optimal solutions for small-scale problems but has scalability limitation. Towards this end, a heuristic algorithm was designed to minimize the number of bad segments, which has linear complexity with respect to the network size.

2.1.3 Others

Apart from IP/port address shuffling and network topology mutation, there are some other MTD perturbation strategies. Li et al. [13] designed a novel network traffic morphing algorithm to ensure that the network sessions within cyber-physical systems are indistinguishable from other typical network sessions from the statistic view of inter-packet delays. The data package will be sent through one of these network sessions with assured real-time requirement, where the morphing process is dynamically adjusted with low overhead and complexity.

2.2 Physical Moving Target Defense

Although the MTD scheme is originated from the computer science community, numerous meaningful and interesting research works are surging to introduce the novel MTD idea into the smart grid area for security enhancement. In particular, the increasing digitization and intelligence of sensor, controller, and actuator have brought new possibility for implementing MTD strategies in the smart grid. This section aims to survey physical MTD approaches in the smart grid.

2.2.1 Moving Target Defense for Power System State Estimation

Morrow et al. [19, 20] pioneered the proactive probing strategy against the stealthy FDI attack by randomizing the power line admittance through D-FACTS devices in the power grid. The D-FACTS perturbation command is kept secret from the attacker through a "key space" method. Following this research line, Rahman et al. [21] formally named this D-FACTS based proactive defense as MTD and supplemented the perturbation of measurement set used for SE as a new MTD scheme. To reveal the trade-off between detection enhancement and operation alteration under MTD, Liu et al. [33] applied the rank of composite matrix to maximize the detection likelihood while minimizing the power loss of transmission lines. In addition to the well-recognized attack detection enhancement benefiting from MTD, Tian et al. [23] investigated a hiddenness-enhanced MTD strategy such that the detection effectiveness can be improved with enhanced hiddenness. It is proven that MTD hiddenenss is equivalent to maintaining the power flow, thus revealing that complete detection and hiddenness are two conflicting goals.

To counter against the CCPA that masks the impact of physically disconnected power lines by tampering with cyber measurements [24], Lakshminarayana et al. [25] applies MTD to invalidate the knowledge used for crafting CCPA vectors. In order to minimize the defense cost during the operational stage, a zero-sum game is formulated to identify the subset of power lines that need to be installed with D-FACTS as well as the perturbation magnitude to deter the strategic and intelligent adversary from causing catastrophe consequences. Furthermore, Lakshminarayana et al. [26] continued to investigate the relationship between the MTD effectiveness in detecting FDI attacks and associated growth of operational cost, where the *smallest principal angle* between the column spaces of measurement matrices before and after the admittance perturbation is used to trade off the MTD properties.

Regarding the detection and identification of FDI attacks, Zhang et al. [27] provided sufficient and necessary conditions in terms of MTD design. Moreover, a heuristics algorithm was presented for the deployment of D-FACTS such that the number of vulnerable state variables after MTD can be reduced as much as possible. To improve the attack detection capability, coordinated perturbation in consecutive intervals was devised as a feasible option. Considering the impact of uncertain noises, Xu et al. [28] designed the robust MTD that is able to maximize the worst-case detection rate against FDI attacks while limiting the attack probability on undetectable subspace. To further improve the detection effectiveness, Zhang et al. [29] proposed the EMTD by coordinately perturbing the series reactance and parallel susceptance of power lines, and designed the cost-minimization EMTD considering detection improvement as well as the locational marginal price, power loss, and device cost.

Based on the graph-theory-topology analysis, Liu et al. [30, 31] analyzed sufficient conditions for maintaining MTD effectiveness and hiddenness, based on which the MTD planning and operation that maximizes detection effectiveness and guarantees hiddenness are devised. Considering the stability impact of admittance perturbation, Zhang et al. [31] derived a voltage stability indicator with respect to admittance perturbation and integrated it into the MTD design for the balance of detection effectiveness and stability requirements. To further improve the state recovery capability, Liu et al. [34] tried to integrate meter protection with admittance perturbation, derived conditions for recovering all system states, and presented a heuristic algorithm for deploying and implementing the MTD strategy.

2.2.2 Moving Target Defense for Microgrid Control

In addition to the MTD strategy in the power grid SE, Liu et al. [35, 36] proposed a PDDL framework to disclose stealthy FDI attacks in the microgrid. As shown in Fig. 2.4, the PDDL consists of detection and localization components: (1) In each upper host, two attack detection indicators, i.e., VBD and CSD, are calculated to perceive the attack impact caused by stealthy FDI attacks in microgrids; (2) Once the detection alarm is raised by upper host, the perturbation on generation unit's control gains will be activated to quickly and accurately localize the compromised generation units. The control gain perturbation is optimized to maximize the localization capability while ensuring negligible transient fluctuation on the system states. After anomaly detection, recursive impact mitigation strategies [44, 45] have also been developed to recover the legitimate data.

By introducing uncertainty into the communication channels between the data transmitter and receiver, Giraldo et al. [37] devised a DMTDR framework as shown in Fig. 2.5. The DMTDR integrates two layers of uncertainty, namely random replica activation and path selection: (1) The random replica activation layer selects a random subset of replicas to transmit their information at a given time; (2) The path selection algorithm selects which one of the transmitted replicas reaches their intended destination. These two layers are able to make it harder for adversaries to learn details about the system while reducing the impact and success probability of FDI attacks.

Different from perturbing system configurations such as the control gain and communication channel, Li et al. [38, 39] proposed a proactive attack detection method that is able to locate the infracted controller components by injecting programmable small probe signals into the control input. Besides injecting watermarks into the control channel, Liu et al. [40, 41] validated the effectiveness of embedding noise-

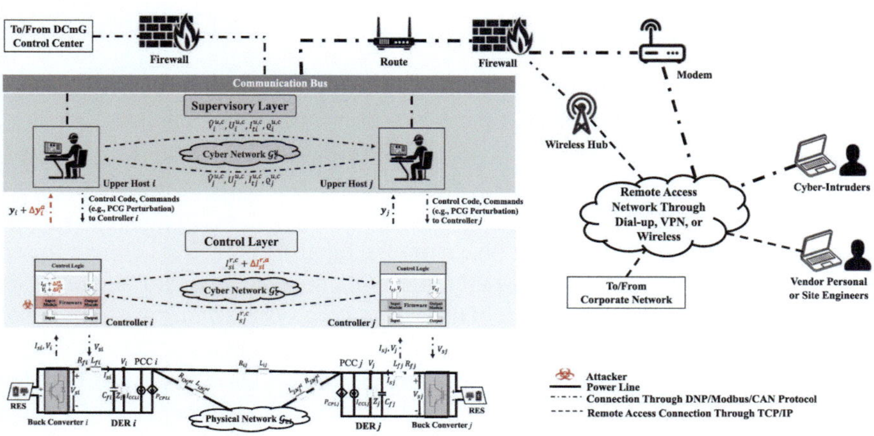

Fig. 2.4 The implementation of PDDL framework in microgrids [35, 36]

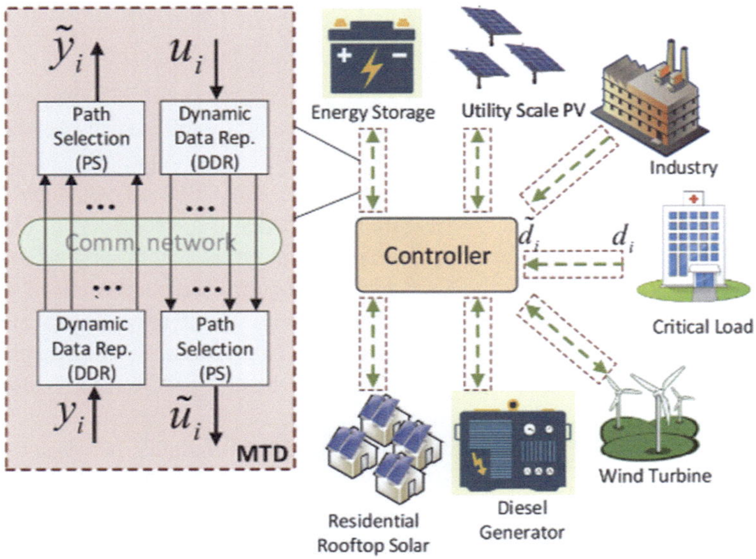

Fig. 2.5 The implementation of DMTDR framework in microgrids [37]

equivalent watermarking signals into the sensor readings, where the watermarking strength is optimized to maximize the detection enhancement while avoiding control performance degradation.

2.2.3 *Moving Target Defense Analysis from Control Theory Perspective*

Starting from the control perspective, Kanellopoulos et al. [16] presented a secure control framework to harmonically integrate proactive and reactive defense mechanisms. The proactive mechanism is originated from the MTD scheme to dynamically and constantly alert the system parameters in a stochastic switching structure such that the unpredictability of utilized sensors and actuators is maximized in the reconnaissance stage. The reactive phase, on the other hand, detects compromised components by leveraging online data through computing an integral Bellman error. The stability of the equilibrium point of the closed-loop system under the adoption of secure control framework is guaranteed with bounded performance loss. To defend against the Stuxnet-like attack that has strong stealthiness, Tian et al. [22] proposed to ingeniously perturb the sensor and actuator configurations and investigated analytical conditions under which this type of attack could be detected.

2.3 Cyber-Physical Coordinated Moving Target Defense

The research work that is able to integrate both cyber and physical attributes into the MTD design is rare, and only one representative work is demonstrated as follows.

By randomizing communications and crafting decoy data for virtual nodes, Lin et al. [42, 43] proposed DefRec to greatly prolong the reconnaissance stage and mislead adversaries into designing damage-free attacks. As shown in Fig. 2.6, the core component of DefRec is the PFV that "hooks" network interactions with real physical devices and mimics thee real devices to create virtual nodes. To build lightweight virtual nodes that follow the implementation of network stack, system invariants, and physical state variations of real devices, the PFV consists of three main componets: (1) virtual node templates contain basic configuration of the target control networks such as the IP space; (2) seed devices profile the runtime behaviors of devices from different vendors including the network stack, system invariant, and physical state; (3) packet hooking, enabled by the programmability of SDN controller, tailors the network packets from seed devices based on virtual node templates and device profiles as the network flow from virtual nodes.

2.4 Summary

This chapter reviews the existing MTD development trend in the context of smart grid. The reviewed MTD strategies are divided into the cyber and physical ones considering the origination of uncertainties that are used by MTD to thwart and harden the host scanning and reconnaissance stage, as well as disclose malicious activities. In particular, the cyber MTD takes IP address/port shuffling and network topology mutation into account, while the physical MTD considers the scenarios of power grid SE and microgrid control as well as the theory analysis perspective. Finally, the cyber-physical coordinated MTD that integrates cross-domain attributes is discussed in details.

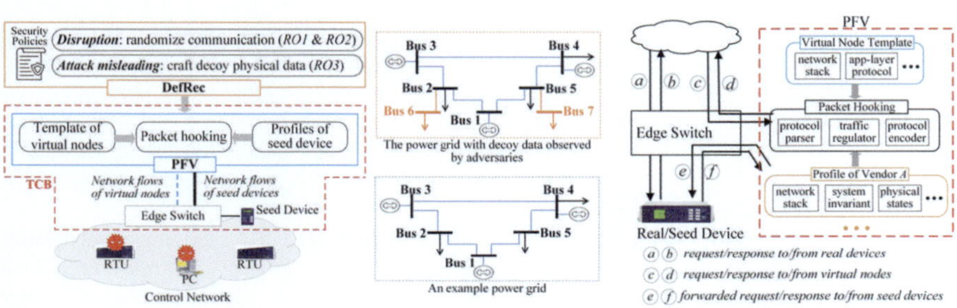

Fig. 2.6 The implementation framework of DefRec [42, 43]

References

1. D.P. Sharma, D.S. Kim, S. Yoon, H. Lim, J.-H. Cho, T.J. Moore, FRVM: flexible random virtual IP multiplexing in software-defined networks, in *Proceedings of the 17th IEEE International Conference on Trust, Security and Privacy in Computing and Communications/12th IEEE International Conference on Big Data Science and Engineering (TrustCom/BigDataSE)* (2018), pp. 579–587
2. D.C. MacFarland, C.A. Shue, The SDN shuffle: creating a moving-target defense using host-based software-defined networking, in *Proceedings of the Second ACM Workshop on Moving Target Defense* (2015), pp. 37–41
3. S. Antonatos, P. Akritidis, E.P. Markatos, K.G. Anagnostakis, Defending against hitlist worms using network address space randomization, in *Proceedings of the 2005 ACM Workshop on Rapid Malcode* (2005), pp. 30–40
4. J.H. Jafarian, E. Al-Shaer, Q. Duan, Openflow random host mutation: transparent moving target defense using software-defined networking, in *Proceedings of the First Workshop on Hot Topics in Software-Defined Networks* (2012), pp. 127–132
5. T.E. Carroll, M. Crouse, E.W. Fulp, K.S. Berenhaut, Analysis of network address shuffling as a moving target defense, in *Proceedings of the 2014 IEEE International Conference on Communications (ICC)* (2014), pp. 701–706
6. P. Kampanakis, H. Perros, and T. Beyene, SDN-based solutions for moving target defense network protection, in *Proceedings of the IEEE International Symposium on a World of Wireless, Mobile and Multimedia Networks (WoWMoM) 2014* (2014), pp. 1–6
7. S. Sengupta, A. Chowdhary, A. Sabur, A. Alshamrani, D. Huang, S. Kambhampati, A survey of moving target defenses for network security. IEEE Commun. Surv. & Tutor. **22**(3), 1909–1941 (2020)
8. J.-H. Cho, D.P. Sharma, H. Alavizadeh, S. Yoon, N. Ben-Asher, T.J. Moore, D.S. Kim, H. Lim, F.F. Nelson, Toward proactive, adaptive defense: a survey on moving target defense. IEEE Commun. Surv. & Tutor. **22**(1), 709–745 (2020)
9. S. Achleitner, T.F. La Porta, P. McDaniel, S. Sugrim, S.V. Krishnamurthy, R. Chadha, Deceiving network reconnaissance using SDN-based virtual topologies. IEEE Trans. Netw. Serv. Manag. **14**(4), 1098–1112 (2017)
10. J.B. Hong, S. Yoon, H. Lim, D.S. Kim, Optimal network reconfiguration for software-defined networks using shuffle-based online MTD, in *Proceedings of the 2017 IEEE 36th Symposium on Reliable Distributed Systems (SRDS)* (2017), pp. 234–243
11. Y.-B. Luo, B.-S. Wang, G.-L. Cai, Effectiveness of port hopping as a moving target defense, in *Proceedings of the 2014 7th International Conference on Security Technology* (2014), pp. 7–10
12. A. Chowdhary, A. Alshamrani, D. Huang, H. Liang, MTD analysis and evaluation framework in software defined network (MASON), in *Proceedings of the 2018 ACM International Workshop on Security in Software Defined Networks & Network Function Virtualization* (2018), pp. 43–48
13. Y. Li, R. Dai, J. Zhang, Morphing communications of cyber-physical systems towards moving-target defense, in *Proceedings of the 2014 IEEE International Conference on Communications (ICC)* (2014), pp. 592–598
14. S. Groat, M. Dunlop, W. Urbanski, R. Marchany, J. Tront, Using an IPv6 moving target defense to protect the smart grid, in *Proceedings of the 2012 IEEE PES Innovative Smart Grid Technologies (ISGT)* (2012), pp. 1–7
15. T. Chin, K. Xiong, Dynamic generation containment systems (DGCS): a moving target defense approach, in *Proceedings of the 2016 3rd International Workshop on Emerging Ideas and Trends in Engineering of Cyber-Physical Systems (EITEC)* (2016), pp. 11–16. https://doi.org/10.1109/EITEC.2016.7503690.
16. A. Kanellopoulos, K.G. Vamvoudakis, A moving target defense control framework for cyber-physical systems. IEEE Trans. Autom. Control **65**(3), 1029–1043 (2019)

17. A.C. Pappa, A. Ashok, M. Govindarasu, Moving target defense for securing smart grid communications: architecture, implementation & evaluation, in *Proceedings of the 2017 IEEE Power & Energy Society Innovative Smart Grid Technologies Conference (ISGT)* (2017), pp. 1–5
18. J. Ulrich, J. Drahos, M. Govindarasu, A symmetric address translation approach for a network layer moving target defense to secure power grid networks, in *Proceedings of the 2017 Resilience Week (RWS)* (2017), pp. 163–169
19. K.L. Morrow, E. Heine, K.M. Rogers, R.B. Bobba, T.J. Overbye, Topology perturbation for detecting malicious data injection, in *Proceedings of the 2012 45th Hawaii International Conference on System Sciences (HICSS)* (2012), pp. 2104–2113
20. K.R. Davis, K.L. Morrow, R. Bobba, E. Heine, Power flow cyber attacks and perturbation-based defense, in *Proceedings of the 2012 IEEE Third International Conference on Smart Grid Communications (SmartGridComm)* (2012), pp. 342–347
21. M.A. Rahman, E. Al-Shaer, R.B. Bobba, Moving target defense for hardening the security of the power system state estimation, in *Proceedings of the First ACM Workshop on Moving Target Defense* (2014), pp. 59–68
22. J. Tian, R. Tan, X. Guan, Z. Xu, T. Liu, Moving target defense approach to detecting Stuxnet-like attacks. IEEE Trans. Smart Grid **11**(1), 291–300 (2019)
23. J. Tian, R. Tan, X. Guan, T. Liu, Enhanced hidden moving target defense in smart grids. IEEE Trans. Smart Grid **10**(2), 2208–2223 (2019). https://doi.org/10.1109/TSG.2018.2791512
24. R. Deng, P. Zhuang, H. Liang, CCPA: coordinated cyber-physical attacks and countermeasures in smart grid. IEEE Trans. Smart Grid **8**(5), 2420–2430 (2017). https://doi.org/10.1109/TSG.2017.2702125
25. S. Lakshminarayana, E.V. Belmega, H.V. Poor, Moving-target defense against cyber-physical attacks in power grids via game theory. IEEE Trans. Smart Grid **12**(6), 5244–5257 (2021)
26. S. Lakshminarayana, D.K.Y. Yau, Cost-benefit analysis of moving-target defense in power grids. IEEE Trans. Power Syst. **36**(2), 1152–1163 (2020)
27. Z. Zhang, R. Deng, P. Cheng, M.-Y. Chow, Strategic protection against FDI attacks with moving target defense in power grids. IEEE Trans. Control Netw. Syst. **9**(1), 245–256 (2021)
28. W. Xu, I.M. Jaimoukha, F. Teng, Robust moving target defence against false data injection attacks in power grids. IEEE Trans. Inf. Forens. Secur. **18**, 29–40 (2023). https://doi.org/10.1109/TIFS.2022.3210864
29. M. Zhang, X. Fan, R. Lu, C. Shen, X. Guan, Extended moving target defense for AC state estimation in smart grids. IEEE Trans. Smart Grid **14**(3), 2313–2325 (2022)
30. B. Liu, H. Wu, Optimal D-FACTS placement in moving target defense against false data injection attacks. IEEE Trans. Smart Grid **11**(5), 4345–4357 (2020). https://doi.org/10.1109/TSG.2020.2977207
31. H. Zhang, B. Liu, X. Liu, A. Pahwa, H. Wu, Voltage stability constrained moving target defense against net load redistribution attacks. IEEE Trans. Smart Grid **13**(5), 3748–3759 (2022). https://doi.org/10.1109/TSG.2022.3170839
32. B. Liu, H. Wu, Optimal planning and operation of hidden moving target defense for maximal detection effectiveness. IEEE Trans. Smart Grid **12**(5), 4447–4459 (2021). https://doi.org/10.1109/TSG.2021.3076824
33. C. Liu, J. Wu, C. Long, D. Kundur, Reactance perturbation for detecting and identifying FDI attacks in power system state estimation. IEEE J. Select. Topics Signal Proc. **12**(4), 763–776 (2018). https://doi.org/10.1109/JSTSP.2018.2846542
34. C. Liu, H. Liang, T. Chen, J. Wu, C. Long, Joint admittance perturbation and meter protection for mitigating stealthy FDI attacks against power system state estimation. IEEE Trans. Power Syst. **35**(2), 1468–1478 (2020). https://doi.org/10.1109/TPWRS.2019.2938223
35. M. Liu, C. Zhao, J. Xia, R. Deng, P. Cheng, J. Chen, PDDL: proactive distributed detection and localization against stealthy deception attacks in DC microgrids. IEEE Trans. Smart Grid **14**(1), 714–731 (2022)
36. M. Liu, C. Zhao, Z. Zhang, R. Deng, P. Cheng, J. Chen, Converter-based moving target defense against deception attacks in DC microgrids. IEEE Trans. Smart Grid **13**(5), 3984–3996 (2022)

37. J. Giraldo, M. El Hariri, M. Parvania, Decentralized moving target defense for microgrid protection against false-data injection attacks. IEEE Trans. Smart Grid **13**(5), 3700–3710 (2022)
38. Y. Li, P. Zhang, L. Zhang, B. Wang, Active synchronous detection of deception attacks in microgrid control systems. IEEE Trans. Smart Grid **8**(1), 373–375 (2016)
39. Y. Li, Y. Qin, P. Zhang, A. Herzberg, SDN-enabled cyber-physical security in networked microgrids. IEEE Trans. Sustain. Energy **10**(3), 1613–1622 (2018)
40. M. Liu, X. Zhang, H. Zhu, Z. Zhang, R. Deng, Physics-aware watermarking embedded in unknown input observers for false data injection attack detection in cyber-physical microgrids. IEEE Trans. Inf. Forens. Secur. **19**, 7824–7840 (2024). https://doi.org/10.1109/TIFS.2024.3447235
41. H. Zhu, M. Liu, C. Fang, R. Deng, P. Cheng, Detection-performance tradeoff for watermarking in industrial control systems. IEEE Trans. Inf. Forens. Secur. **18**, 2780–2793 (2023). https://doi.org/10.1109/TIFS.2023.3269919
42. H. Lin, J. Zhuang, Y.-C. Hu, H. Zhou, DefRec: establishing physical function virtualization to disrupt reconnaissance of power grids' cyber-physical infrastructures, In *Proceedings of the 2020 Network and Distributed System Security Symposium (NDSS)* (2020)
43. H. Lin, Z.T. Kalbarczyk, R.K. Iyer, RAINCOAT: andomization of network communication in power grid cyber infrastructure to mislead attackers. IEEE Trans. Smart Grid **10**(5), 4893–4906 (2018)
44. M. Liu, X. Zhang, C. Zhao, R. Deng, Matrix coding enabled impact mitigation against primary false data injection attacks in cyber-physical microgrids. IEEE Trans. Power Syst. **1**(1), 1–16 (2025). https://doi.org/10.1109/TPWRS.2025.3528322
45. M. Liu, X. Zhang, R. Zhang, Z. Zhou, Z. Zhang, R. Deng, Detection-triggered recursive impact mitigation against secondary false data injection attacks in cyber-physical microgrid. IEEE Trans. Smart Grid **16**(2), 1744–1761 (2025). https://doi.org/10.1109/TSG.2024.3493754

Chapter 3
Design of Moving Target Defense in the Smart Grid

3.1 System Model and Threat Model

Throughout this chapter, calligraphy font ($\mathcal{L}$) indicates a set or a subspace ($\mathcal{A}$), math boldface font ($\mathbf{H}$) indicates a matrix, bold lower case letter ($\mathbf{x}$) indicates an vector. $S(\cdot)$ denotes the (column) span of a matrix. $R(\cdot)$ is the rank of a matrix. $\cdot^T$ indicates the transpose of a matrix. $|\cdot|$ denotes the cardinality of a set.

3.1.1 System Model

To avoid intensive computation and obtain an optimal solution for large power systems, power system engineers usually utilize a linearized DC power flow model to approximate the AC power flow model [1, 47]. For its computational speed and simplicity, the DC model has been widely used for decades in both industry and academia [2, 13, 38, 42]. Although it is less accurate, the DC model is more robust and often used in real-time operations such as the computation of marginal price [26]. In the DC model, the voltage magnitude is by default set as 1 p.u. The state variables are reduced to the voltage phase angle. The measurements are reduced to active power flows. Moreover, since the phase angle difference is usually constrained to be small, the power flow equations can be reduced to

$$f_{ij} = -b_{ij}(\theta_i - \theta_j), \quad p_i = \sum_{j \in \mathcal{K}_i} f_{ij}, \tag{3.1}$$

where f_{ij} indicates the active power flow between bus i and bus j, b_{ij} is the equivalent susceptance on branch $\{i, j\}$, θ_i and θ_j are respectively the voltage phase angle of bus i and j, p_i is the active power injection of bus i, $\mathcal{K}_i$ is a set of neighboring buses connected to bus i.

R. Deng et al., *Moving Target Defense in the Smart Grid*, Wireless Networks,
https://doi.org/10.1007/978-3-031-92378-4_3

We consider a classic power transmission network consisting of a set $\mathcal{N} = \{1, 2, \ldots, n+1\}$ of buses and a set $\mathcal{L} = \{k_1, k_2, \ldots, k_l\}$ of branches, where $n+1$ is the number of buses and l is the number of branches. With the DC model, by setting an arbitrary bus as the reference/slack bus, the remaining n phase angles $\theta_1, \theta_2, \ldots, \theta_n$ are taken to form the system states, typically denoted as $\mathbf{x} \in \mathbb{R}^n$. Each branch $k_t = \{i, j\} \in \mathcal{L}$ connects two buses i and j. Assuming that the power system is fully measured, i.e., each bus is monitored by one meter and each branch is monitored by two meters (both in the positive and negative direction). Then, the number of meter measurements is $m = 2l + n + 1$. The DC model can be derived as

$$\mathbf{z} = \mathbf{H}\mathbf{x} + \boldsymbol{\eta}, \tag{3.2}$$

where $\mathbf{H}$ is termed as the measurement matrix, $\mathbf{z}$ denotes the measurements of active power injections and active power flows, $\boldsymbol{\eta}$ represents the independent measurement noises, which are typically assumed to be normally distributed [i.e., $\eta_i \sim \mathcal{N}(0, \sigma_i^2)$].

Construction of measurement matrix. Let $\mathbf{A}$ denote the branch-bus incidence matrix. It is given by

$$a_{ti} = \begin{cases} 1, & \text{if branch } k_t \text{ starts from bus } i; \\ -1, & \text{if branch } k_t \text{ ends at bus } i; \\ 0, & \text{otherwise,} \end{cases} \tag{3.3}$$

where a_{ti} is the element in the position (t, i) of the matrix $\mathbf{A}$. Let $\mathbf{D}$ denote the diagonal branch susceptance matrix. Its diagonal element $\mathbf{D}_{tt}$ is $-b_{ij}$ with branch $k_t = \{i, j\}$. Thus, the invertible symmetric admittance matrix is $\mathbf{B} = \mathbf{A}^T\mathbf{D}\mathbf{A}$ and the branch-bus shift factor matrix is $\mathbf{S} = \mathbf{D}\mathbf{A}$. Considering the DC power flow equations, we have $\mathbf{f} = \mathbf{S}\boldsymbol{\theta}$ and $\mathbf{p} = \mathbf{B}\boldsymbol{\theta}$, where $\mathbf{f}$ denotes the active power flows, $\mathbf{p}$ denotes the power injections, $\boldsymbol{\theta}$ denotes the phase angles. Therefore, the measurement matrix $\mathbf{H} \in \mathbb{R}^{m \times n}$ can be derived as

$$\mathbf{H} = \begin{bmatrix} \mathbf{B} \\ \mathbf{S} \\ -\mathbf{S} \end{bmatrix} = \begin{bmatrix} \mathbf{A}^T\mathbf{D}\mathbf{A} \\ \mathbf{D}\mathbf{A} \\ -\mathbf{D}\mathbf{A} \end{bmatrix}. \tag{3.4}$$

We can see that the measurement matrix $\mathbf{H} \in \mathbb{R}^{m \times n}$ is a function of the system topology and branch susceptances.

In cases that the power systems are partially measured (i.e., some buses and branches are not monitored by meters), the corresponding measurement matrix is formed by selecting some rows from the measurement matrix of the fully measured case. For any measurement matrix, we assume that $m > n$ in this chapter. Note that all results in this chapter are satisfied whenever the system is fully measured or partially measured.

State estimation. State estimator, a fundamental tool for economically and dynamically routing power flows, is responsible to optimally estimate the state variables with noisy measurements collected by the underlying SCADA system [47]. The most common and concise mathematical description of SE is

$$\hat{\mathbf{x}} = (\mathbf{H}^T\mathbf{W}\mathbf{H})^{-1}\mathbf{H}^T\mathbf{W}\mathbf{z} = \mathbf{K}\mathbf{z}, \tag{3.5}$$

where $\mathbf{W}$ is a diagonal matrix whose elements are reciprocals of the noise variances, $\mathbf{K}$ is referred to as the "pseudo-inverse" of $\mathbf{H}$ because $\mathbf{KH} = \mathbf{I}$. $\hat{\mathbf{x}}$ is calculated by using certain statistical estimation criterion such as maximum likelihood, weighted least-squares, etc.

Bad data detection. Normally, in order to filter out abnormal/erronuous meter measurements, the BDD checker is used in SE. Let $\mathbf{r} = \mathbf{z} - \mathbf{H}\hat{\mathbf{x}} = (\mathbf{I} - \mathbf{HK})\mathbf{z}$ be the residues between the measurements $\mathbf{z}$ and their estimates $\hat{\mathbf{z}} = \mathbf{H}\hat{\mathbf{x}}$. BDD compares the Euclidean norm $\|\mathbf{r}\|_2$ with a predetermined threshold τ. If $\|\mathbf{r}\|_2 > \tau$, the abnormal alarm is triggered; otherwise, the measurements $\mathbf{z}$ are considered normal.

In the following, we assume that the measurements are noiseless (i.e., $\mathbf{e} = \mathbf{0}$) to simplify the discussions. Under this assumption, the DC model can be written as $\mathbf{z} = \mathbf{H}\mathbf{x}$. The corresponding state estimates are given by $\hat{\mathbf{x}} = (\mathbf{H}^T\mathbf{H})^{-1}\mathbf{H}^T\mathbf{z} = \mathbf{x}$. Moreover, the threshold τ for the BDD is equal to 0. Nevertheless, the main results in the following derivations still keep valid with noisy measurements. In our simulations (see Sect. 3.4), we will evaluate the impacts of measurement noises.

3.1.2 *Threat Model*

As a critical infrastructure, the security issue of power grid has attracted a lot of attention. For example, the US NESCOR records safety incidents and negative impacts on the physical objects in power systems [36]. NERC reports lessons learned from failures and blackouts caused by cyber system faults [35]. These recordings show the security risks of power systems, which should be well addressed during daily maintenance and operation.

Adversarial setting. In this chapter, we consider the class of stealthy FDI attacks studied in [29]. Let $\mathbf{a} \in \mathbb{R}^m$ be the attack vector. The malicious measurements are given by $\mathbf{z}_a = \mathbf{z} + \mathbf{a}$. It has been proved that $\mathbf{z}_a$ cannot be detected by the BDD if $\mathbf{a} = \mathbf{Hc}$ [29], where $\mathbf{c} \in \mathbb{R}^n$ is an arbitrary vector. That is, $\mathbf{r}^a = \mathbf{z}^a - \mathbf{H}\hat{\mathbf{x}}^a = \mathbf{z} + \mathbf{a} - \mathbf{H}(\hat{\mathbf{x}} + \mathbf{c}) = \mathbf{z} - \mathbf{H}\hat{\mathbf{x}} = \mathbf{r}$, where $\mathbf{r}^a$ denotes the measurement residues after the FDI attack and $\hat{\mathbf{x}}^a$ denotes the modified states. We can see that the measurement residues are not changed after the FDI attack. Thus, the bad data alarm is not triggered while the system states are subverted.

In other words, let $\mathcal{A} \in \mathbb{R}^m$ be a subspace that contains all FDI attacks with the form of $\mathbf{a} = \mathbf{Hc}$. Then, $\mathcal{A}$ is given by

$$\mathcal{A} = \{\mathbf{a} \mid \mathbf{a} = \mathbf{Hc},\ \ \mathbf{c} \in \mathbb{R}^n\}. \tag{3.6}$$

We can see that $\mathcal{A}$ is equal to the subspace $S(\mathbf{H})$, where $S(\cdot)$ denotes the (column) span of a matrix. Therefore, if $\mathbf{a} \in S(\mathbf{H})$ holds, then the malicious measurements $\mathbf{z}_a$ can bypass the BDD. In this chapter, we assume the attacker has the following capabilities:

- The attacker is able to know a measurement matrix $\mathbf{H}$ based on his/her understanding, e.g., through topology-leaking attacks [57] or subspace attacks [19]. Since these topology attacks depend on historical measurements, the measurement matrix $\mathbf{H}$ cannot be learned by the attacker immediately. The learning/inferring process usually takes a sufficiently long time (hours or days) due to the exfiltration of an enough amount of historical measurement data [5, 19, 23, 24, 50].
- The attacker is able to eavesdrop and tamper with the measurements by intruding into the communication network or IP-accessible field devices [41]. Practically, the attacker will have limited resources to compromise all the measurements [29]. However, we do not assume a priori what specific data can be compromised or not. Note also that although the attacker might be able to compromise the confidentiality of the raw data points, he/she will need non-trivial efforts and time to learn their higher system level relationships to guide his attack.
- The attacker is unable to take over (or have full access to) the control center or the SCADA systems. Once the attacker can exploit the control center or the SCADA systems, he/she is powerful enough to thwart most defensive mechanisms.

3.2 Problem Statement

Based on the system model and threat model, in this section, we first introduce the approach of MTD used in power grid. Then, we present the problems mainly investigated in this chapter.

3.2.1 Moving Target Defense by Perturbing Branch Susceptances

D-FACTS. The D-FACTS devices are first introduced by Divan and Johal for controlling the power flow [11]. These devices can alter the impedances of power lines, and thus, control power flows to eliminate transmission constraints and bottlenecks. They are small and light enough to be suspended from the power line, floating both electrically and mechanically on it. Moreover, the equipped communication system enables them to receive control commands and transmit working states to remote control stations [17]. To date, a lot of researches have devoted to analyzing the performance of D-FACTS devices and investigating the use of them in different power system applications [4, 15, 32, 39, 46].

Moving target defense. Recently, some pioneer works have exploited the adoption of D-FACTS devices for thwarting FDI attacks in power grid [7, 20, 24, 27, 33, 34, 45, 53]. Morrow et al. [7, 34] were the first researchers who proposed to perturb branch impedances for probing both the malicious and bad data in the power grid. The following works [24, 45, 53] named this branch perturbation strategy as MTD. Here we briefly introduce this defensive mechanism based on the DC model.

With D-FACTS devices, the defender is able to actively perturb a set of branch susceptances, an thus, increase system uncertainty for potential attackers. Supposing the susceptance of branch $k_t = \{i, j\}$ is perturbed, then we have

$$b_{ij} \to b'_{ij}, \tag{3.7}$$

where b'_{ij} is the susceptance of branch k_t after MTD. Actually, we cannot change b_{ij} as much as we can [34, 39, 45]. There are limits on the perturbed result, i.e.,

$$b_{ij}^{min} \leq b'_{ij} \leq b_{ij}^{max}. \tag{3.8}$$

As a result, the measurement matrix is changed. We assume that the control commands of MTD can be protected in the control center and transmitted through safeguarded communication channels. Thus, the attacker is unable to anticipate them. We use $\mathbf{H}'$ to denote the new measurement matrix after MTD. In fact, the attacker might try to learn the perturbed measurement matrix $\mathbf{H}'$. But the learning process usually takes a sufficiently long time (hours or days) due to the exfiltration of an enough amount of historical measurement data [5, 19, 23, 24, 50]. In other words, the attacker cannot obtain the latest measurement matrix immediately. Thus, we can dynamically change the measurement matrix accordingly before it risks being exposed. The execution cycle of MTD has been discussed by authors in [24, 45].

Since the current susceptance perturbations are not anticipated by the attacker, he/she only knows the former measurement matrix $\mathbf{H}$ (may not be the measurement matrix just before MTD). Thus, if the attacker still constructs the attack vector as $\mathbf{a} = \mathbf{Hc}$, it is possible that $\mathbf{a} \notin S(\mathbf{H}')$ after MTD. In other words, the malicious measurements $\mathbf{z}_a = \mathbf{z}' + \mathbf{Hc}$ may not bypass the BDD with $\mathbf{H}'$, where $\mathbf{z}'$ are the measurements after MTD. The defender's BDD after MTD is given by

$$r(\mathbf{z}') = \|\mathbf{z}' - \mathbf{H}'\mathbf{K}'\mathbf{z}'\|, \tag{3.9}$$

where $\mathbf{K}' = (\mathbf{H}'^T\mathbf{H}')^{-1}\mathbf{H}'^T$. Since $m > n$, we can prove that $\mathbf{H}'\mathbf{K}' \neq \mathbf{I}$. Therefore, $r(\mathbf{z}') = 0$ if and only if $\mathbf{z}' \in S(\mathbf{H}')$ with noiseless measurements [29]. Since $\mathbf{z}' \in S(\mathbf{H}')$, $r(\mathbf{z}_a) = 0$ if and only if $\mathbf{a} \in S(\mathbf{H}')$. This means that attack vector can bypass the BDD after MTD if and only if it belongs to the following *stealthy attack space*.

Definition 3.1 *Let $\mathcal{A}_s$ denote the stealthy attack space. Then, $\mathcal{A}_s = S(\mathbf{H}) \cap S(\mathbf{H}')$, i.e.,*

$$\mathcal{A}_s = \{\mathbf{a} \mid \mathbf{a} = \mathbf{Hc} \wedge \mathbf{a} = \mathbf{H}'\mathbf{c}', \ \ \mathbf{c}, \mathbf{c}' \in \mathbb{R}^n\}. \tag{3.10}$$

We can see that the stealthy attack space is the intersection of the column space of $\mathbf{H}$ and $\mathbf{H}'$. Therefore, the stealthy attack space is also a subspace and its dimension is given by

$$\begin{aligned} & dim(\mathcal{A}_s) \\ & = dim(S(\mathbf{H}) \cap S(\mathbf{H}')) \\ & = dim(S(\mathbf{H})) + dim(S(\mathbf{H}')) - dim\big(S(\mathbf{H}) \cup S(\mathbf{H}')\big) \\ & = 2n - R([\mathbf{H}\ \mathbf{H}']), \end{aligned} \tag{3.11}$$

where $dim(\cdot)$ is the dimension of a subspace, $R(\cdot)$ is the rank of a matrix. We can see that the dimension of the stealthy attack space $\mathcal{A}_s$ is closely related to the rank of the combined matrix $[\mathbf{H}\ \mathbf{H}']$, i.e., $R([\mathbf{H}\ \mathbf{H}'])$. We define $R([\mathbf{H}\ \mathbf{H}'])$ as the security factor of MTD and denote it as γ.

3.2.2 Problem Statement

In this chapter, we mainly analyze the effectiveness of MTD from two aspects: (i) the capability of MTD to protect the system from any FDI attack constructed with the former measurement matrix, and (ii) if the first capability cannot be achieved, the capability of MTD to narrow the attack opportunities of potential attackers by appropriately deploying the D-FACTS devices and setting the susceptance perturbation magnitudes.

As we know, as long as the stealthy attack space $\mathcal{A}_s$ is not $\{\mathbf{0}\}$, it is still possible for the attacker to successfully execute FDI attacks after MTD. The intuitive understanding is that, if $\exists\, \mathbf{c}, \mathbf{c}' \in \mathbb{R}^n$ satisfying $\mathbf{a} = \mathbf{Hc} = \mathbf{H}'\mathbf{c}' \neq \mathbf{0}$ [i.e., $\mathbf{a} \in \mathcal{A}_s$], then the malicious measurements $\mathbf{z}_a = \mathbf{z}' + \mathbf{a}$ after MTD can circumvent the defender's BDD. We take the 4-bus power system (Fig. 3.1) as an example. Suppose that the branch susceptances known by the attacker are $b_{12} = -1, b_{13} = -2, b_{14} = -3, b_{23} = -4$ and $b_{34} = -5$. If we perturb the branches $\{1, 2\}$ and $\{2, 3\}$ as $\Delta b_{12} = 0.1$ and $\Delta b_{13} = 0.2$, the attacker still can construct an attack vector as $\mathbf{a} = [0, 0, 3c, 0, 5c]^T$ to bypass the BDD after MTD, where $c \in \mathbb{R}$ is the error injected into the state variable of bus 4. Here the dimension of the stealthy attack space is 1. In other words, the MTD may

Fig. 3.1 A 4-bus power system with 5 branches

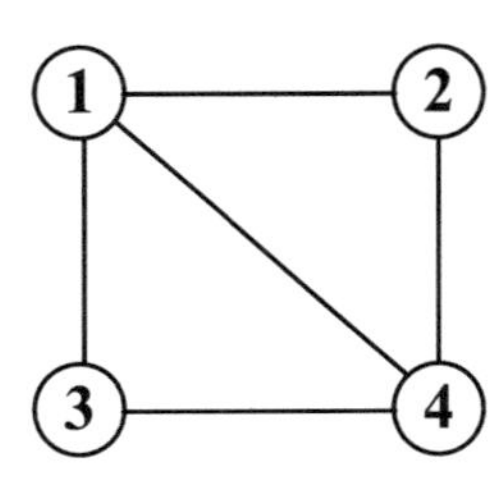

Table 3.1 The dimension of the stealthy attack space after MTD

Branch perturbations	$dim(\mathcal{A}_s)$
$\Delta b_{12} = 0.1$, $\Delta b_{13} = 0.1$	1
$\Delta b_{12} = 0.1$, $\Delta b_{23} = 0.1$	2
$\Delta b_{13} = 0.1$, $\Delta b_{14} = 0.3$, $\Delta b_{34} = 0.5$	2
$\Delta b_{13} = 0.1$, $\Delta b_{14} = 0.3$, $\Delta b_{34} = 0.2$	1

not be complete to thwart all FDI attacks with the form of $\mathbf{a} = \mathbf{Hc}$. We define the MTD's completeness as follows.

Definition 3.2 *An MTD is said to be complete if and only if all attack vectors in $\mathcal{A}$ except the zero vector cannot bypass the BDD after MTD. This indicates that $\mathcal{A}_s = \{\mathbf{0}\}$.*

Besides, once a complete MTD cannot be achieved, we need to exploit how to reduce the stealthy attack space as much as possible, thus, reducing the probability of stealthy FDI attacks after MTD. An intuitive idea is to perturb all branches in the power network. But it is an unrealistic assumption to deploy D-FACTS on all branches. On the other hand, we cannot randomly choose a set of target-perturbation branches and arbitrarily perturbed them. The impacts on the stealthy attack space might be different if we perturb different branches and set different perturbation values. For example, given the 4-bus power system shown in Fig. 3.1, Table 3.1 shows the change of the dimension of the attack space when we perturb different branches and set different perturbation values. From row 1 and row 2, we can see that the dimensions of the stealthy attack space are different when we perturb different branches. Form row 3 and row 4, we can see that the dimensions of the stealthy attack space are different when we perturb branch {3, 4} to different values.

Moreover, due to the physical limitation of D-FACTS devices [4] and their induced operation costs [24, 45], the branch susceptance cannot be perturbed to any value. Therefore, a natural question emerges that, given a set of D-FACTS devices, how do they affect the stealthy attack space and the operation cost when we deploy them on different branches and set them with different values? This is a critical issue for us to carry out MTD.

3.3 Analysis of Moving Target Defense to Thwart False Data Injection Attacks

In this section, first, we analyze the detection of FDI attacks constructed with the former measurement matrix using MTD. Second, we analyze the completeness of MTD and discuss physical limitations of the power network to achieve this property. Third, we investigate the impact of the susceptance perturbation magnitude on the dimension of the stealthy attack space. Lastly, we provide guidance on effective MTD for minimizing the dimension of the stealthy attack space, maximizing the number of covered buses, and reducing the operation cost.

3.3.1 Detecting False Data Injection Attacks with Moving Target Defense

First of all, we discuss the detection of FDI attacks constructed with the former measurement matrix.

After MTD, from the measurement matrix $\mathbf{H}$, we select u columns which form a submatrix $\mathbf{H}^u$ such that $\mathbf{H}^u$ can be linearly represented by $\mathbf{H}'$. The rest v ($v = n - u$) columns in $\mathbf{H}$ form a submatrix $\mathbf{H}^v$, i.e., $\mathbf{H} = [\mathbf{H}^u\ \mathbf{H}^v]$. Assume that the attacker constructs an attack vector as $\mathbf{a} = \mathbf{Hc}$ with the purpose to bypass the BDD after MTD. Actually, $\mathbf{a} = \mathbf{H}^u\mathbf{c}_u + \mathbf{H}^v\mathbf{c}_v$, where $\mathbf{c}_u \in \mathbb{R}^u$ and $\mathbf{c}_v \in \mathbb{R}^v$. Then, we have the following conclusion.

Proposition 3.1 The BDD after MTD can detect the attack vector $\boldsymbol{a} = \mathbf{H}\boldsymbol{c}$ only if $\boldsymbol{c}_v \neq \boldsymbol{0}_v$, where $\boldsymbol{0}_v$ is a v-dimension zero vector.

Proof Since $\mathbf{H}^u$ can be linearly represented by $\mathbf{H}'$, we have $\mathbf{H}^u\mathbf{c}_u \in S(\mathbf{H}')$ for any $\mathbf{c}_u \in \mathbb{R}^u$. Suppose $\mathbf{c}_v = \mathbf{0}_v$. Then, we have $\mathbf{a} = \mathbf{H}^u\mathbf{c}_u + \mathbf{H}^v\mathbf{c}_v \in S(\mathbf{H}')$ (i.e., $\mathbf{a} \in \mathcal{A}_s$), which means that we cannot detect this FDI attack. Therefore, we must have $\mathbf{c}_v \neq \mathbf{0}_v$ for detecting the attack vector $\mathbf{a}$.

Proposition 3.1 presents a necessary condition for the detection of FDI attacks constructed with the former measurement matrix. It also indicates the soundness of MTD. That is, if the BDD after MTD can detect the attack vector $\mathbf{a}$, then at least one of the state variables corresponding to $\mathbf{c}_v$ is modified. With this condition, if $\mathbf{H}^v$ cannot be linearly represented by $\mathbf{H}'$, then there might not exist $\mathbf{c}' \in \mathbb{R}^n$ such that $\mathbf{H}^v\mathbf{c}_v = \mathbf{H}'\mathbf{c}'$ for an arbitrarily selected $\mathbf{c}_v$. Thus, at most v state variables corresponding to $\mathbf{c}_v$ cannot be independently modified by the attacker. Since $v \geq n - dim(\mathcal{A}_s)$, if the dimension of the stealthy attack space $dim(\mathcal{A}_s)$ is smaller, the more state variables cannot be independently modified by the attacker after MTD. However, note that the attackers cannot anticipate/predict the new measurement matrix $\mathbf{H}'$, the "blind" attacker might try his/her luck to execute FDI attacks, and thus, increase the detection probability of these attacks with the BDD after MTD.

Specifically, to illustrate the effectiveness of MTD for thwarting FDI attacks, we give an illustration example (Fig. 3.2) about the change of the stealthy attack space with respect to the security factor γ. We can see that, the stealthy attack space is equal to the column space of $\mathbf{H}$ and $\mathbf{H}'$ [$S(\mathbf{H}) = S(\mathbf{H}')$] when the security factor γ is n; the stealthy attack space is the intersection of the column space of $\mathbf{H}$ and $\mathbf{H}'$ [$S(\mathbf{H}) \neq S(\mathbf{H}')$] when the security factor γ is between n and $2n$; the stealthy attack space is $\{\mathbf{0}\}$ when the security factor γ is $2n$. Therefore, if the security factor γ is smaller, the "volume" of the stealthy attack space is smaller. A smaller stealthy

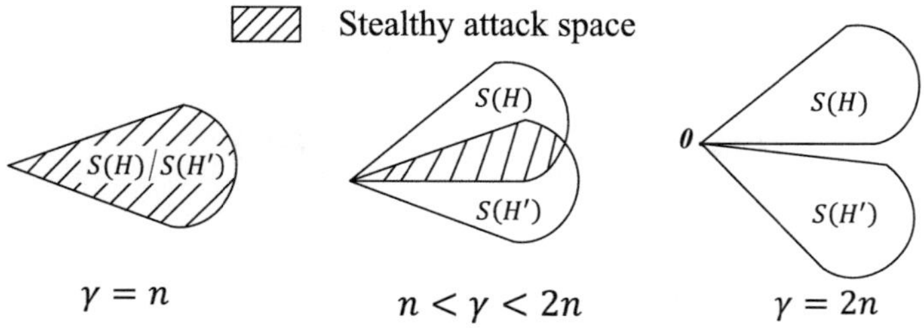

Fig. 3.2 An illustration example about the change of the stealthy attack space $\mathcal{A}_s$ with the security factor γ

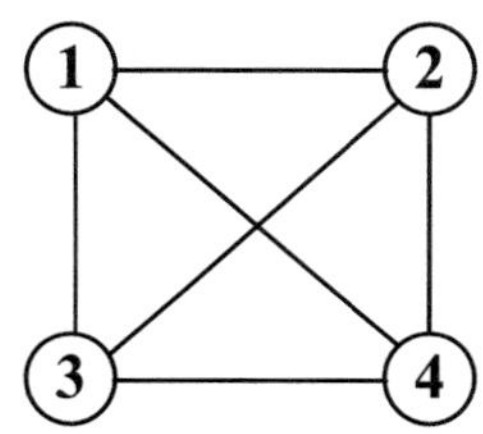

Fig. 3.3 A 4-bus power system with 6 branches

attack space means a smaller probability of stealthy FDI attacks constructed with the former measurement matrix. In other words, after MTD, there are less successful opportunities for the attacker to execute stealthy FDI attacks with the old system information. Thus, we claim that MTD is more effective to thwart stealthy FDI attacks. For example, when $\gamma = 2n$, after MTD, no attack vector $\mathbf{a} = \mathbf{Hc}$ can bypass BDD, which means that the probability of stealthy FDI attacks constructed with the former measurement matrix is zero. We will evaluate the effectiveness of MTD with respect to the dimension of the stealthy attack space in Sect. 3.4.1.

Considering the 4-bus power system shown in Fig. 3.3, the branch susceptances known by the attacker are $b_{12} = -1$, $b_{13} = -2$, $b_{14} = -3$, $b_{23} = -4$, $b_{24} = -5$ and $b_{34} = -6$. If we do not perturb any branches, then all attack vectors $\mathbf{a} = \mathbf{Hc}$ can bypass the BDD (i.e., $\gamma = n = 3$, $dim(\mathcal{A}_s) = 3$); If we perturb the susceptance of branch $\{1, 2\}$ to $b'_{12} = -1.1$, then we can reduce the dimension of the stealthy attack space to 2 (i.e., $3 < \gamma = 4 < 6$, $dim(\mathcal{A}_s) = 2$), but there still exist attack vector $\mathbf{a} = \mathbf{Hc}$ that can bypass the BDD after MTD, e.g., $\mathbf{c} = [0, c_2, c_3]^T, c_2, c_3 \in \mathbb{R}$; If we perturb the susceptances of branches $\{1, 2\}$, $\{1, 3\}$ and $\{2, 4\}$ as $b'_{12} = -1.1$, $b'_{13} = -2.15$, $b'_{24} = -5.1$, then any FDI attack constructed as $\mathbf{a} = \mathbf{Hc}$ ($\mathbf{a} \neq \mathbf{0}$) cannot bypass the BDD after MTD (i.e., $\gamma = 2n = 6$, $dim(\mathcal{A}_s) = 0$). That is, this MTD is complete. We find that we can realize a complete MTD only if more than 3 branches are perturbed under this 4-bus power system with 6 branches. The completeness of MTD is investigated in the next subsection.

3.3.2 Analysis of Completeness of Moving Target Defense

Here we first give a sufficient and necessary condition for achieving a complete MTD mathematically. Then, we present physical constraints for achieving this property.

Mathematical Analysis

In fact, an MTD is not complete because the BDD after MTD misses detecting some FDI attacks, i.e., $\mathcal{A}_s \neq \{\mathbf{0}\}$. According to the Definition 3.2, to achieve a complete MTD, we need to make $\mathcal{A}_s = \{\mathbf{0}\}$. Since the subspace spanned by the zero vector has zero dimension, the dimension of $\mathcal{A}_s$ is zero if an MTD is complete. Thus, we have the following conclusion.

Proposition 3.2 An MTD is complete if and only if the security factor $\gamma = 2n$.

Proof (Sufficiency) Since $\gamma = R([\mathbf{H}\,\mathbf{H}']) = 2n$, we have $dim(\mathcal{A}_s) = dim(S(\mathbf{H}) \cap S(\mathbf{H}')) = 0$. It follows that $\mathcal{A}_S = \{\mathbf{0}\}$. Therefore, the MTD is complete.

(Necessity) Since the MTD is complete, we have $\mathcal{A}_s = \{\mathbf{0}\}$. It follows that $dim(\mathcal{A}_s) = dim(S(\mathbf{H}) \cap S(\mathbf{H}')) = 2n - R([\mathbf{H} \quad \mathbf{H}']) = 0$. Thus, we have $\gamma = R([\mathbf{H}\,\mathbf{H}']) = 2n$.

Proposition 3.2 presents a sufficient and necessary condition for achieving a complete MTD mathematically. Note that if $\gamma = R([\mathbf{H}\,\mathbf{H}']) = 2n$, the intersection of $S(\mathbf{H})$ and $S(\mathbf{H}')$ is $\{\mathbf{0}\}$. Thus, all FDI attacks constructed as $\mathbf{a} = \mathbf{Hc}$ $(\mathbf{a} \neq \mathbf{0})$ can be detected by the BDD after MTD.

Remark 3.1 Moreover, if an MTD is complete, then we can identify the attack vector constructed with the former measurement matrix. Suppose the attack vector is $\mathbf{a} = \mathbf{Hc}$. Then, the malicious measurements are $\mathbf{z}_a = \mathbf{H}'\mathbf{x}' + \mathbf{Hc} = [\mathbf{H}'\,\mathbf{H}][\mathbf{x}'\,\mathbf{c}]^T$, where $\mathbf{x}'$ denotes the state variables without FDI attacks. Since the combined matrix $[\mathbf{H}'\,\mathbf{H}]$ has full column rank, we can uniquely solve the variables $\mathbf{c}$ with the malicious measurements. Thus, we can identify the injected errors of the state variables.

Physical Limitations

Essentially, whether we can construct a complete MTD or not depends on the structure of the power network and the perturbed branches. Considering the fully measured case, the measurement matrix before and after MTD are

$$\mathbf{H} = \begin{bmatrix} \mathbf{A}^T\mathbf{DA} \\ \mathbf{DA} \\ -\mathbf{DA} \end{bmatrix} \qquad \mathbf{H}' = \begin{bmatrix} \mathbf{A}^T\mathbf{D}'\mathbf{A} \\ \mathbf{D}'\mathbf{A} \\ -\mathbf{D}'\mathbf{A} \end{bmatrix}. \tag{3.12}$$

The combined matrix of $\mathbf{H}$ and $\mathbf{H}'$ can be written as $\mathbf{C} = [\mathbf{H}\,\mathbf{H}']$. Let $\widetilde{\mathbf{S}} = [\mathbf{S}\;\mathbf{S}'] = [\mathbf{DA}\;\mathbf{D}'\mathbf{A}]$, we have

$$\mathbf{C} = \begin{bmatrix} \mathbf{A}^T\widetilde{\mathbf{S}} \\ \widetilde{\mathbf{S}} \\ -\widetilde{\mathbf{S}} \end{bmatrix} = \begin{bmatrix} \mathbf{A}^T & \mathbf{0}_{n\times l} & \mathbf{0}_{n\times l} \\ \mathbf{0}_{l\times l} & \mathbf{I}_{l\times l} & \mathbf{0}_{l\times l} \\ \mathbf{0}_{l\times l} & \mathbf{0}_{l\times l} & -\mathbf{I}_{l\times l} \end{bmatrix} \begin{bmatrix} \widetilde{\mathbf{S}} \\ \widetilde{\mathbf{S}} \\ \widetilde{\mathbf{S}} \end{bmatrix}, \tag{3.13}$$

where $\mathbf{0}_{n\times l}$ is an n by l zero matrix, $\mathbf{I}_{l\times l}$ is an l by l identity matrix. Therefore, we can derive that $R(\mathbf{C}) \leq R(\widetilde{\mathbf{S}})$. Since $\widetilde{\mathbf{S}}$ is an l by $2n$ matrix, we have $R(\widetilde{\mathbf{S}}) \leq min\{l, 2n\}$. Therefore, only if $l \geq 2n$ can we make $R(\mathbf{C}) = 2n$. In other words, in order to make the security factor $\gamma = R(\mathbf{C}) = 2n$, the number of branches of the power transmission system must be larger than or equal to $2n$. We observe that this condition must be satisfied under the other measured cases (i.e., the system is not necessarily fully measured) as well.

Theorem 3.1 An MTD is complete only if the following two conditions are satisfied:

- *$l \geq 2n$, where l is the number of branches and $n + 1$ is the number of buses in the power system.*
- The perturbed branches must cover all buses.

Proof Considering the first condition, we resort to a contradiction. That is, suppose that the MTD is complete but $l < 2n$. Let $\widehat{\mathbf{H}}$ and $\widehat{\mathbf{H}}'$ denote the measurement matrices before and after MTD with a fully measured power system, respectively. Thus, with any partially measured power system, the measurement matrix $\mathbf{H}$ is formed by selecting some rows from $\widehat{\mathbf{H}}$. Therefore, the combined matrix $[\mathbf{H}\;\mathbf{H}']$ can be formed by selecting some rows from the combined matrix $[\widehat{\mathbf{H}}\;\widehat{\mathbf{H}}']$. It follows that $R([\mathbf{H}\;\mathbf{H}']) \leq R([\widehat{\mathbf{H}}\;\widehat{\mathbf{H}}'])$. We have proved that $R([\widehat{\mathbf{H}}\;\widehat{\mathbf{H}}']) = 2n$ only if the number of branches l is greater than or equal to $2n$. Therefore, we can derive that $R([\mathbf{H}\;\mathbf{H}']) \leq R([\widehat{\mathbf{H}}\;\widehat{\mathbf{H}}']) < 2n$ if $l < 2n$. This indicates that the MTD is not complete, which contradicts to the original assumption. Therefore, we must guarantee $l \geq 2n$ for achieving a complete MTD.

Next, we consider the second condition. Let $\mathcal{M}_{mtd}$ be the set of buses covered by the perturbed branches and n_s is the size of $\mathcal{M}_{mtd}$. $\mathbf{H}^q$ is a submatrix formed by selecting q columns from $\mathbf{H}'$ such that the combined matrix $[\mathbf{H}\;\mathbf{H}^q]$ has full column rank. Let $\mathcal{M}^q_{mtd}$ be an index set of these q columns in $\mathbf{H}'$. Since $\mathcal{M}^q_{mtd} \subseteq \mathcal{M}_{mtd}$, we have $q \leq n_s$. If the MTD is complete, then we have $q = n$. Therefore, $n_s \geq n$, that is, the perturbed branches must cover all buses.

Theorem 3.1 provides a necessary condition for realizing a complete MTD. That is, the completeness of MTD can be achieved only if the system topology and the perturbed branches meet certain requirements. Besides, we can derive another two points from the second condition. First, to cover all buses, at least n branches should be perturbed. Second, once a bus is not covered by the perturbed branches, the state variable of this bus can be modified stealthily if the attacker happens to attack it only. That is, let $\mathbf{a} = \mathbf{Hc}$ be an attack vector. Then, if $c_i = 0$ ($c_i \in \mathbf{c}$) for any $i \in \mathcal{M}_{mtd}$, then the malicious measurements $\mathbf{z}_a = \mathbf{z}' + \mathbf{a}$ can bypass the BDD after MTD, where $\mathcal{M}_{mtd}$ is a non-empty set of buses that are covered by the perturbed branches. Thus, $c_i \neq 0$ ($i \notin \mathcal{M}_{mtd}$) can be any values. This indicates that the attacker can arbitrarily modify the state variables of buses that are not covered by the perturbed branches. Therefore, for a complete MTD, we need to ensure that: (i) the power transmission system has more than $2n$ branches; (ii) more than n branches are perturbed; (iii) the perturbed branches cover all buses. For example, the 4-bus power system shown in Fig. 3.3 can support the realization of a complete MTD, because it has 6 branches ($2n = 2 * 3 = 6$), while the 4-bus power system with 5 branches shown in Fig. 3.1 cannot. If any one of the above three conditions is not satisfied, we cannot make an MTD complete.

Considering the first condition, if the power system has less than $2n$ branches, then the dimension of the stealthy attack space satisfies

$$dim(\mathcal{A}_s) = 2n - R([\mathbf{H}\ \mathbf{H}']) \geq 2n - R(\widetilde{\mathbf{S}}) \geq 2n - l. \tag{3.14}$$

That is, the smallest dimension of the stealthy attack space after MTD is larger than or equal to $2n - l$.

Remark 3.2 *According to the aforementioned analysis, we find that is difficult to realize a complete MTD. For a complete MTD, the power transmission system must have at least* $2n$ *branches (i.e.,* $l \geq 2n$*), and more than n branches, which cover all buses, should be perturbed. These conditions may not be held in practice. Particularly, we examine the number of branches in the IEEE test power systems provided in MATPOWER* [58]*; only three of all 41 cases have more than* $2n$ *branches, namely case6ww (11 branches), case89pegase (210 branches) and case145 (453 branches). What's more, we discover that we can never make an MTD complete if the the power transmission system has a bus that is only connected by a single branch.*

Theorem 3.2 It is impossible to make an MTD complete if the power transmission system contains a bus that is only connected by a single branch.

Proof Suppose t is the bus that is only connected by a single branch, the susceptance of the connected branch $k_d = \{t', t\}$ is $b_{t't}$, $\mathbf{e}_i \in \{0, 1\}^n$ is a vector with an unit in the ith position and zero elsewhere, and $\mathbf{u}_{ij} = \mathbf{e}_i - \mathbf{e}_j$ ($k_d = \{i, j\} \in \mathcal{L}$) is a vector with "1" in the ith position and "–1" in the jth position and zero elsewhere. Considering the fully measured case, let the tth column of the branch-bus shift factor matrix $\mathbf{S}$ and the symmetric admittance matrix $\mathbf{B}$ be $\mathbf{s}_t$ and $\mathbf{b}_t$, respectively. Then, we have $\mathbf{s}_t = -b_{t't}\mathbf{e}_t$ and $\mathbf{b}_t = b_{t't}\mathbf{u}_{t't}$. Suppose $\mathbf{s}'_t = -b'_{t't}\mathbf{e}_t$ and $\mathbf{b}'_t = b'_{t't}\mathbf{u}_{t't}$ after MTD. Then, we can derive that $\mathbf{s}_t = \frac{b_{t't}}{b'_{t't}}\mathbf{s}'_t$ and $\mathbf{b}_t = \frac{b_{t't}}{b'_{t't}}\mathbf{b}'_t$. Let $\mathbf{h}_t$ and $\mathbf{h}'_t$ be the tth column of $\mathbf{H}$ and $\mathbf{H}'$ before and after MTD, respectively. Then, we have $\mathbf{h}_t = \frac{b_{t't}}{b'_{t't}}\mathbf{h}'_t$. Since any $\mathbf{H}$ can be formed by selecting some rows from the measurement matrix under the fully measured case, we can derive $\mathbf{h}_t = \frac{b_{t't}}{b'_{t't}}\mathbf{h}'_t$ under the partially measured case as well.

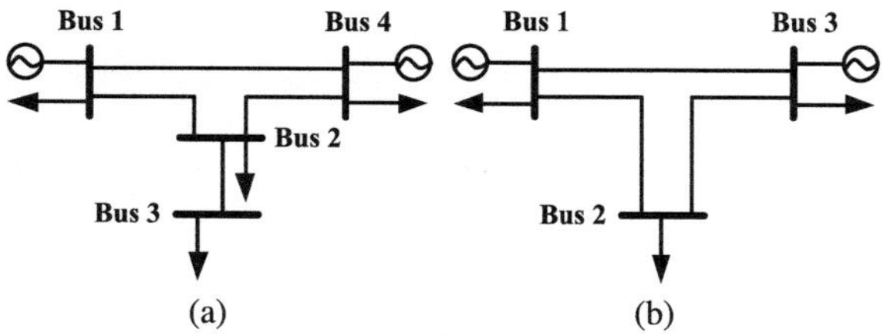

Fig. 3.4 **a** A 4-bus power system with 4 branches; **b** A 3-bus power system

It follows that $S(\mathbf{H}') = S(\mathbf{H})$ if we only perturb the branch k_d. Therefore, we can never obtain $\gamma = R([\mathbf{H}\ \mathbf{H}']) = 2n$.

This special case of the power network structure limits the realization of a complete MTD. Besides, we find that the state variable of the bus that is only connected by a single branch can be arbitrarily modified by the attacker. Let t be the bus that is only connected by a single branch. Then, for any $c_t \in \mathbb{R}$, there exists $c'_t \in \mathbb{R}$ such that $\mathbf{h}_t c_t = \mathbf{h}'_t c'_t$, where $\mathbf{h}_t$ and $\mathbf{h}'_t$ are respective the tth column of the matrix $\mathbf{H}$ and $\mathbf{H}'$. Therefore, if the attack vector is $\mathbf{a} = \mathbf{h}_t c_t$, then $\mathbf{a} \in S(\mathbf{H}')$ always holds, which can definitely bypass the BDD after MTD. We present a simple example to illustrate that. In the 4-bus power system shown in Fig. 3.4, the state variable of bus 3 can be arbitrarily modified because it is only connected by branch $\{2, 3\}$. Therefore, as long as the system contains a bus that is only connected by a single branch, we can never realize a complete MTD.

Discussion

Nevertheless, it might not be necessary to achieve a complete MTD in some power systems. We take the 3-bus power system shown in Fig. 3.4 as an example. Even though we cannot realize a complete MTD because it only has 3 branches, which is less than $2 * 2 = 4$, we can protect all state variables from being arbitrarily modified by the attacker when we perturb the branches $\{1, 2\}$ and $\{2, 3\}$. Suppose the perturbations are Δb_{12} and Δb_{23}, and the attack vector is $\mathbf{a} = \mathbf{Hc}$, where $\mathbf{c} = [c_1\ c_2]^T$. If this attack can bypass the BDD after MTD, c_1 and c_2 must satisfy

$$\frac{c_2}{c_1} = \frac{b_{12}\Delta b_{23} - b_{23}\Delta b_{12}}{b_{12}\Delta b_{23} + \Delta b_{12}\Delta b_{23}}. \tag{3.15}$$

Therefore, the attacker must know the susceptance perturbations Δb_{12} and Δb_{23} to construct a coordinated attack vector $\mathbf{a}$. We can see that is almost impossible for the attacker to construct such an attack vector. Definitely, a complete MTD can thwart any attack vector $\mathbf{a} = \mathbf{Hc}$ when $\mathbf{H}$ is very different from the measurement matrix $\mathbf{H}'$ after MTD. But it might not be a necessary condition for some power systems.

3.3.3 Impact of Susceptance Perturbation Magnitude on Dimension of Stealthy Attack Space

A practical and effective way to enhance the security of the power system is to reduce the stealthy attack space $\mathcal{A}_s$ and cover as many buses as possible with MTD, thus reducing the attack opportunities of potential attackers. We start by investigating the impact of the susceptance perturbation magnitude on the dimension of the stealthy attack space when one more branch susceptance is perturbed. Since the security factor γ [i.e., $R([\mathbf{H}\ \mathbf{H}'])$] determines $dim(\mathcal{A}_s)$, we focus on the change of γ in the following. As $\mathbf{H}' = \mathbf{H} + \Delta\mathbf{H}$, based on the sparse property of $\Delta\mathbf{H}$, we first present three cases about the change of γ when one more branch is perturbed. Second, we investigate the impact of the susceptance perturbation magnitude on the value of γ with these three cases.

Before go deeper into the analysis, we prove that $\Delta\mathbf{H}$ has a sparse structure.

Proposition 3.3 Suppose $\mathcal{L}_d$ is a set of perturbed branches. $\Delta\mathbf{H}$ is a sparse matrix with non-zero elements in the i_dth and j_dth columns, with $k_d = \{i_d, j_d\} \in \mathcal{L}_d$.

Proof Let $\mathbf{e}_i \in \{0, 1\}^n$ be a vector with a unit in the ith position and zeros elsewhere, and let $\mathbf{u}_{ij} = \mathbf{e}_i - \mathbf{e}_j$ ($t_k = \{i, j\} \in \mathcal{L}$) be a vector with "1" in the ith position and "–1" in the jth position and zeros elsewhere. Then, we can rewrite $\mathbf{A}$ and $\mathbf{D}$ as

$$\mathbf{A} = \sum_{\substack{t_k \in \mathcal{L} \\ t_k = \{i,j\}}} \mathbf{e}_k \mathbf{u}_{ij}^T , \qquad \mathbf{D} = \sum_{\substack{t_k \in \mathcal{L} \\ t_k = \{i,j\}}} -b_{ij} \mathbf{e}_k \mathbf{e}_k^T . \tag{3.16}$$

Supposing the diagonal branch susceptance matrix after MTD is $\mathbf{D}'$ and $\Delta\mathbf{D} = \mathbf{D}' - \mathbf{D}$, then we can derive $\Delta\mathbf{D}$ as

$$\begin{aligned}\Delta\mathbf{D} &= \sum_{\substack{k_d \in \mathcal{L}_d \\ k_d = \{i_d, j_d\}}} -(b'_{i_d j_d} - b_{i_d j_d}) \mathbf{e}_d \mathbf{e}_d^T \\ &= \sum_{\substack{k_d \in \mathcal{L}_d \\ k_d = \{i_d, j_d\}}} -\Delta b_{i_d j_d} \mathbf{e}_d \mathbf{e}_d^T .\end{aligned} \tag{3.17}$$

We can see that there are non-zero elements in positions (k_d, k_d) with $k_d \in \mathcal{L}_d$ and zeros elsewhere in $\Delta\mathbf{D}$. Since $\mathbf{e}_i^T \mathbf{e}_j = 0$ if $i \neq j$ and $\mathbf{e}_i^T \mathbf{e}_j = 1$ if $i = j$, we can derive that the change of the branch-bus shift factor matrix is

$$\begin{aligned}\Delta\mathbf{S} = \Delta\mathbf{D}\mathbf{A} &= \sum_{\substack{k_d \in \mathcal{L}_d \\ k_d = \{i_d, j_d\}}} -(\Delta b_{i_d j_d} \mathbf{e}_d \mathbf{e}_d^T)(\mathbf{e}_d \mathbf{u}_{i_d j_d}^T) \\ &= \sum_{\substack{k_d \in \mathcal{L}_d \\ k_d = \{i_d, j_d\}}} -\Delta b_{i_d j_d} \mathbf{e}_d \mathbf{u}_{i_d j_d}^T\end{aligned} \tag{3.18}$$

We can observe that $\Delta\mathbf{S}$ is a sparse matrix with the dth row is non-zero and other rows are all zero. That is, if $k_d \in \mathcal{L}_d$, then the dth row of $\Delta\mathbf{S}$ is $\Delta\mathbf{S}_d = [\cdots - \Delta b_{i_d j_d} \cdots \Delta b_{i_d j_d} \cdots]$ with $-\Delta b_{i_d j_d}$ in the i_dth column and $\Delta b_{i_d j_d}$ in the j_dth column; if $k_d \notin \mathcal{L}_d$, $\Delta\mathbf{S}_d = [0 \cdots 0 \cdots 0]$. Similarly, since $\mathbf{B} = \mathbf{A}^T\mathbf{S}$, we have

$$\begin{aligned}\Delta\mathbf{B} = \mathbf{A}^T\Delta\mathbf{S} &= \sum_{\substack{k_d \in \mathcal{L}_d \\ k_d = \{i_d, j_d\}}} -\Delta b_{i_d j_d}(\mathbf{u}_{i_d j_d}\mathbf{e}_d^T)\big(\mathbf{e}_d\mathbf{u}_{i_d j_d}^T\big) \\ &= \sum_{\substack{k_d \in \mathcal{L}_d \\ k_d = \{i_d, j_d\}}} -\Delta b_{i_d j_d}\mathbf{u}_{i_d j_d}\mathbf{u}_{i_d j_d}^T.\end{aligned} \tag{3.19}$$

$\Delta\mathbf{B}$ is a sparse matrix with the i_dth and j_dth columns are non-zero and the other columns are all zero.

On the basis of the measurement matrix $\mathbf{H}$ (before MTD) and $\mathbf{H}'$ (after MTD). Since $\mathbf{H}' = [\mathbf{B} + \Delta\mathbf{B}; \mathbf{S} + \Delta\mathbf{S}; -\mathbf{S} - \Delta\mathbf{S}]$, we can derive that the difference of the measurement matrix is

$$\Delta\mathbf{H} = \mathbf{H}' - \mathbf{H} = \begin{bmatrix} \Delta\mathbf{B} \\ \Delta\mathbf{S} \\ -\Delta\mathbf{S} \end{bmatrix}. \tag{3.20}$$

Obviously, we can see that $\Delta\mathbf{H}$ is a sparse matrix with non-zero elements in the i_dth and j_dth column with $k_d = \{i_d, j_d\} \in \mathcal{L}_d$.

For example, if we only perturb a single branch, that is, $\mathcal{L}_d = \{k_d\}$ with $k_d = \{i_d, j_d\}$, then $\Delta\mathbf{H}$ under a fully measured case is stated as

$$\Delta\mathbf{H} = \begin{bmatrix} \mathbf{0} & \vdots & \mathbf{0} & \vdots & \mathbf{0} & \\ \cdots & -\Delta b_{i_d j_d} & \cdots & \Delta b_{i_d j_d} & \cdots & \}i_d \\ \mathbf{0} & \vdots & \mathbf{0} & \vdots & \mathbf{0} & \\ \cdots & \Delta b_{i_d j_d} & \cdots & -\Delta b_{i_d j_d} & \cdots & \}j_d \\ \mathbf{0} & \vdots & \mathbf{0} & \vdots & \mathbf{0} & \\ \cdots & -\Delta b_{i_d j_d} & \cdots & \Delta b_{i_d j_d} & \cdots & \}n+d \\ \mathbf{0} & \vdots & \mathbf{0} & \vdots & \mathbf{0} & \\ \cdots & \Delta b_{i_d j_d} & \cdots & -\Delta b_{i_d j_d} & \cdots & \}n+l+d \\ \mathbf{0} & \underbrace{\vdots}_{i_d} & \mathbf{0} & \underbrace{\vdots}_{j_d} & \mathbf{0} & \end{bmatrix}, \tag{3.21}$$

where $\Delta b_{i_d j_d}$ is the susceptance perturbation of branch k_d, $\mathbf{0}$ is a zero matrix/vector. We can see that all non-zero elements in $\Delta\mathbf{H}$ are in the i_dth column and the j_dth column. Since any measurement matrix is formed by selecting some rows from the measurement matrix $\mathbf{H}$ under the fully measured case, we can derive that $\Delta\mathbf{H}$ is also a sparse matrix with non-zero elements in the i_dth and j_dth column with $k_d = \{i_d, j_d\} \in \mathcal{L}_d$.

Based on the sparse structure of the matrix $\Delta\mathbf{H}$, we can draw a conclusion about the changing range of γ when one more branch is perturbed. Let $\mathbf{C} = [\mathbf{H}\ \mathbf{H}']$ and $\mathbf{C}' = [\mathbf{H}\ \mathbf{H}'']$ be the combined matrices before and after the perturbation of a new branch, respectively. It should be noted that, supposing the perturbed branch is $t_d = \{i_d, j-d\}$, then the only difference between $\mathbf{H}'$ and $\mathbf{H}''$ is that the i_dth and j_dth column have different elements. We denote $\Delta\gamma = R(\mathbf{C}') - R(\mathbf{C})$ as the change of the security factor γ. Then, we obtain the following result.

Proposition 3.4 When perturbing one more branch, $\Delta\gamma$ changes –1, 0 or 1.

Proof We have $\mathbf{C}' = [\mathbf{H}\ \mathbf{H}'] + [\mathbf{0}_{m\times n}\ \ \Delta\mathbf{H}] = \mathbf{C} + [\mathbf{0}_{m\times n}\ \ \Delta\mathbf{H}]$, where $\mathbf{0}_{m\times n}$ is an m by n zero matrix, $\Delta\mathbf{H}$ is a matrix corresponding to the susceptance perturbation of the new branch. According to Eq. (3.21), $R(\Delta\mathbf{H}) = 1$ when one more branch is perturbed. Since $R(\mathbf{C}') \leq R(\mathbf{C}) + R(\Delta\mathbf{H})$, we have $R(\mathbf{C}') \leq R(\mathbf{C}) + 1$. Conversely, we have $\mathbf{C} = [\mathbf{H}\ \mathbf{H}'] + [\mathbf{0}_{m\times n}\ \ -\Delta\mathbf{H}] = \mathbf{C}' + [\mathbf{0}_{m\times n}\ \ -\Delta\mathbf{H}]$. Thus, we can derive that $R(\mathbf{C}) \leq R(\mathbf{C}') + R(-\Delta\mathbf{H}) \Rightarrow R(\mathbf{C}') \geq R(\mathbf{C}) - 1$. Therefore, we have $R(\mathbf{C}) - 1 \leq R(\mathbf{C}') \leq R(\mathbf{C}) + 1$.

That is, there are three possible changes of the security factor γ when one more branch is perturbed: increasing 1, remaining the same or decreasing 1 (i.e., $\Delta\gamma = 1, 0$ or -1). Thus, a natural question emerges that, whether the value of $\Delta\gamma$ will change with the susceptance perturbation magnitude. After the activation of the MTD, suppose $\mathcal{M}^q_{mtd}$ is an index set of q columns in $\mathbf{H}'$ that form a submatrix $\mathbf{H}^q$ such that $R([\mathbf{H}\ \mathbf{H}^q]) = R([\mathbf{H}\ \mathbf{H}']) = n + q$. The rest columns in $\mathbf{H}'$ form a submatrix $\mathbf{H}^p$, i.e., $\mathbf{H}' = [\mathbf{H}^q\ \mathbf{H}^p]$. We define $\mathcal{M}^q_{mtd}$ as a security set. Next, we assume that there is one more branch being perturbed, which is $k_d = \{i_d, j_d\}$. Its susceptance $b_{i_d j_d}$ is perturbed to be $\lambda b_{i_d j_d}$ with $\lambda > 0$ and $\lambda \neq 1$. We define λ as the perturbation ratio. Overall, there are three cases should be considered for the impacts on $\Delta\gamma$ when perturbing this new branch:

- ***Case 1:*** Neither bus i_d nor j_d are contained in the security set. That is, $i_d \notin \mathcal{M}^q_{mtd}$ and $j_d \notin \mathcal{M}^q_{mtd}$.
- ***Case 2:*** Either bus i_d or j_d is contained in the security set. That is, $i_d \in \mathcal{M}^q_{mtd}$ and $j_d \notin \mathcal{M}^q_{mtd}$, or $i_d \notin \mathcal{M}^q_{mtd}$ and $j_d \notin \mathcal{M}^q_{mtd}$.
- ***Case 3:*** Both bus i_d and j_d are contained in the security set. That is, $i_d \in \mathcal{M}^q_{mtd}$ and $j_d \in \mathcal{M}^q_{mtd}$.

For each case, we analyze the change of $\Delta\gamma$ by varying the perturbation ratio λ. Since there are limits on the susceptance perturbation [39], we assume $\lambda_{min} \leq \lambda \leq \lambda_{max}$. We find that the magnitude of the susceptance perturbation almost does not affect the change of the security factor. Especially, if the security factor increases 1 (i.e., $\Delta\gamma = 1$), this result remains the same regardless of the perturbation magnitude. We present the details of the analysis in the following.

Proposition 3.5 Under **Case 1**, if $i_d \notin \mathcal{M}^q_{mtd}$ and $j_d \notin \mathcal{M}^q_{mtd}$, then $\Delta\gamma$ remains the same regardless of the change of λ.

Proof We denote det(·) as the determinant of a matrix. $\mathbf{a}_i$ is the ith column of a matrix $\mathbf{A}$. Since $i_d \notin \mathcal{M}^q_{mtd}$ and $j_d \notin \mathcal{M}^q_{mtd}$, the i_dth column $\mathbf{h}'_{i_d}$ and the j_dth column $\mathbf{h}'_{j_d}$ of $\mathbf{H}'$ can be linearly represented by the columns in $[\mathbf{H}\ \mathbf{H}^q]$. It follows that $R([\mathbf{H}\ \mathbf{H}^q\ \mathbf{h}'_{i_d}]) = R([\mathbf{H}\ \mathbf{H}^q]) = R([\mathbf{H}\ \mathbf{H}'])$. After the susceptance perturbation of branch k_d, $\mathbf{h}'_{i_d}$ and $\mathbf{h}'_{j_d}$ respectively become $\mathbf{h}''_{i_d}$ and $\mathbf{h}''_{j_d}$ according to the proof of Proposition 3.3. We rearrange the matrix $\mathbf{C}'$ as $\mathbf{C}' = [\mathbf{H}\ \mathbf{H}^q\ \mathbf{h}''_{i_d}\ \mathbf{h}''_{j_d}\ \mathbf{H}''_r]$, where $\mathbf{H}''_r$ is a submatrix containing the columns in $\mathbf{H}^p$ ($\mathbf{H}' = [\mathbf{H}^q\ \mathbf{H}^p]$) that are not changed after the perturbation of branch k_d. Let $\mathbf{C}'' = [\mathbf{H}\ \mathbf{H}^q\ \mathbf{h}'_{i_d}\ \mathbf{h}''_{i_d}\ \mathbf{h}''_{j_d}\ \mathbf{H}''_r]$ be a matrix by filling the column $\mathbf{h}'_{i_d}$ into $\mathbf{C}'$. Then, we can derive that $R(\mathbf{C}'') = R(\mathbf{C}')$.

We calculate $R(\mathbf{C}'')$ through elementary column operations. Since only the element relating to branch k_d is changed in column $\mathbf{h}''_{i_d}$, by subtracting $\mathbf{h}''_{i_d}$ from $\mathbf{h}'_{i_d}$, we have $\Delta\mathbf{h}''_{i_d} = \mathbf{h}''_{i_d} - \mathbf{h}'_{i_d} = [0\ \cdots\ 0\ \pm(\lambda-1)b_{i_d j_d}\ 0\ \cdots\ 0\ \pm(\lambda-1)b_{i_d j_d}\ 0\ \cdots\ 0]^T$, where $(\lambda-1)b_{i_d j_d}$ and $-(\lambda-1)b_{i_d j_d}$ are the only non-zero elements in the vector $\Delta\mathbf{h}''_{i_d}$. The number of non-zero elements in this column depends on the measured case of the power system. But it is less than 4 according to the example given in Eq. (3.21). Using the elementary column operation, we transfer the column $\mathbf{h}''_{i_d}$ to $\Delta\mathbf{h}''_{i_d}$ in $\mathbf{C}''$. Since the other elements of the column $\Delta\mathbf{h}''_{i_d}$ are zero, any other column operations on this column are only related to $(\lambda-1)b_{i_d j_d}$. Therefore, it is equivalent to operate on $(\lambda-1)b_{i_d j_d}$ for elementary column operations for any λ if $\lambda \neq 1$. For example, we perturb the susceptance of branch k_d to $\lambda_1 b_{i_d j_d}$ and $\lambda_2 b_{i_d j_d}$, respectively. Assume that only the branches are monitored by meters and each branch is monitored by one meter. Then, we can derive that

$$\Delta\mathbf{h}''_{i_d} = [0\ \cdots\ 0\ \underbrace{(1-\lambda)b_{i_d j_d}}_{k_d^{\text{th}}\ \text{element}}\ 0\ \cdots\ 0]^T, \tag{3.22}$$

If the required column operation is to multiply $\Delta\mathbf{h}''_{i_d}$ with a constant η, then the result is $\mathbf{h} = \eta\Delta\mathbf{h}''_{i_d}$. We find $\mathbf{h} = \eta[0\ \cdots\ 0\ (1-\lambda_1)b_{i_d j_d}\ 0\ \cdots\ 0]^T = \eta \times \frac{\lambda_1}{\lambda_2}[0\ \cdots\ 0\ (1-\lambda_2)b_{i_d j_d}\ 0\ \cdots\ 0]^T$. That is, we can always obtain the same column operation result $\mathbf{h}$ regardless of the value of λ.

Therefore, the impact of $\Delta\mathbf{h}''_{i_d}$ on $R(\mathbf{C}'')$ is not related to the value of λ. Thus, the impact of the column $\mathbf{h}''_{i_d}$ on $R(\mathbf{C}'')$ is not related to the value of λ. Since $R(\mathbf{C}'') = R(\mathbf{C}')$, the impact of the column $\mathbf{h}''_{i_d}$ on $R(\mathbf{C}')$ is not related to the value of λ. As for the column $\mathbf{h}''_{j_d}$, we can draw the same conclusion by using elementary column operations. That is, the impact of the column $\mathbf{h}''_{j_d}$ on $R(\mathbf{C}')$ is not related to the value of λ. Therefore, $\Delta\gamma = R(\mathbf{C}') - R(\mathbf{C})$ does not change with λ ($\lambda > 0$ and $\lambda \neq 1$).

Proposition 3.5 implies that, under ***Case 1***, the perturbation magnitude does not affect the change of $dim(\mathcal{A}_s)$. Therefore, if we find that the dimension of the stealthy attack space $dim(\mathcal{A}_s)$ decreses 1 when we perturb a branch to a certain value, then this result will not change if we perturb the branch to the other values. That is, we can determine the value of $dim(\mathcal{A}_s)$ by only testing one perturbation ratio under ***Case 1***. Considering the other two cases, we obtain the following result.

Proposition 3.6 Under **Case 2** and **Case 3**, only if there exists a value λ^* $(0 < \lambda_{min} \leq \lambda^* \leq \lambda_{max}, \lambda^* \neq 1)$ and $\lambda = \lambda^*$, we obtain $\Delta\gamma = -1$.

Proof We prove this proposition in ***Case 2*** and ***Case 3***, respectively. We use $\det(\cdot)$ to denote the determinant of a matrix. $\mathbf{a}_i$ is the ith column of a matrix $\mathbf{A}$.

Case 2 $: i_d \in \mathcal{M}^q_{mtd}$ ***and*** $j_d \notin \mathcal{M}^q_{mtd}$ Since $\Delta\gamma = -1$, we have $R([\mathbf{H}\ \mathbf{H}'']) = R([\mathbf{H}\ \mathbf{H}']) - 1$. It follows that we must have $R([\mathbf{H}\ \mathbf{H}^q_r\ \mathbf{h}''_{i_d}]) = R([\mathbf{H}\ \mathbf{H}^q]) - 1$ when $\lambda = \lambda^*$, where $\mathbf{H}^q_r$ is a submatrix formed by deleting the column $\mathbf{h}'_{i_d}$ from $\mathbf{H}^q$, $\mathbf{h}'_{i_d}$ is the i_dth column of $\mathbf{H}'$, $\mathbf{h}''_{i_d}$ is the i_dth column of $\mathbf{H}''$ after the perturbation of branch k_d. If we select any $n+q$ rows that contain the perturbed rows (i.e., there are elements $\lambda b_{i_d j_d}$ and/or $-\lambda b_{i_d j_d}$ in this row) and any $n+q$ columns from the combined matrix $[\mathbf{H}\ \mathbf{H}^q_r\ \mathbf{h}''_{i_d}]$, we can form an $n+q$ by $n+q$ square matrix $\mathbf{Q}$. Using the Leibniz formula, $\det(\mathbf{Q})$ is a linear function of $\lambda b_{i_d j_d}$, that is, $\det(\mathbf{Q}) = ab_{i_d j_d}\lambda + b$, where a and b are values calculated by other elements. Specifically, a and b are different if we choose different rows and columns to form $\mathbf{Q}$. Because the combined matrix $[\mathbf{H}\ \mathbf{H}^q_r\ \mathbf{h}''_{i_d}]$ has full column rank when $\lambda = 1$, there exists $\mathbf{Q}$ such that $\det(\mathbf{Q}) = ab_{i_d j_d} + b \neq 0$. Therefore, a and b cannot be 0 at the same time. If $a = 0$, then $\det(\mathbf{Q}) = ab_{i_d j_d}\lambda + b \neq 0$ for any λ. Therefore, there does not exist λ such that $\det(\mathbf{Q}) = 0$. If $a \neq 0$, then $\det(\mathbf{Q}) = ab_{i_d j_d}\lambda + b = 0$ has an unique solution $\frac{-b}{ab_{i_d j_d}}$. As $\Delta\gamma = -1$ when $\lambda = \lambda^*$, we have $R([\mathbf{H}\ \mathbf{H}^q_r\ \mathbf{h}''_{i_d}]) = n + q - 1$. It follows that, for any $\mathbf{Q}$, we have $\det(\mathbf{Q}) = ab_{i_d j_d}\lambda^* + b = 0$. This indicates that λ^* must be equal to $\frac{-b}{ab_{i_d j_d}}$. Therefore, $\Delta\gamma = -1$ only if there exists λ^* such that $\lambda_{min} \leq \lambda^* \leq \lambda_{max}$ and $\lambda^* \neq 1$ and $\lambda = \lambda^*$.

Case 3 $: i_d \in \mathcal{M}^q_{mtd}$ ***and*** $j_d \in \mathcal{M}^q_{mtd}$ After branch k_d is perturbed, we can rearrange $\mathbf{C}' = [\mathbf{H}\ \mathbf{H}'']$ as $\mathbf{C}' = [\mathbf{H}\ \mathbf{H}^q_r\ \mathbf{h}''_{i_d}\ \mathbf{h}''_{j_d}\ \mathbf{H}^p]$, where $\mathbf{H}^q_r$ is a matrix containing the rest columns of $\mathbf{H}^q$ that are not changed, $\mathbf{h}''_{i_d}$ and $\mathbf{h}''_{j_d}$ are the i_dth column and j_dth column in $\mathbf{H}''$ that are perturbed. If we select any $n+q$ rows that contain the perturbed rows (i.e., there are elements $\lambda b_{i_d j_d}$ and/or $-\lambda b_{i_d j_d}$ in this row) and any $n+q$ columns from the combined matrix $[\mathbf{H}\ \mathbf{H}^q_r\ \mathbf{h}''_{i_d}\ \mathbf{h}''_{j_d}]$, we can from an $n+q$ by $n+q$ square matrix $\mathbf{Q}$. Using the Leibniz formula, the determinant of $\mathbf{Q}$ is $\det(\mathbf{Q}) = ab_{i_d j_d}\lambda + b$, where a and b are coefficients after the calculation. As there exists $\mathbf{Q}$ such that we have $\det(\mathbf{Q}) = ab_{i_d j_d} + b \neq 0$ when $\lambda = 1$, a and b cannot be 0 at the same time. Then, if $a = 0$, there exists $\mathbf{Q}$ such that $\det(\mathbf{Q}) \neq 0$ for any λ. Therefore, $\Delta\gamma \neq -1$ for any λ. If $a \neq 0$, $\det(\mathbf{Q}) = ab_{i_d j_d}\lambda + b = 0$ has an unique solution $\frac{-b}{ab_{i_d j_d}}$. Thus, if $R(\mathbf{Q}) = n + q - 1$ (i.e., $\Delta\gamma = -1$) when $\lambda = \lambda^*$, then λ^* must be equal to $\frac{-b}{ab_{i_d j_d}}$. Therefore, $\Delta\gamma = -1$ only if there exists λ^* such that $\lambda_{min} \leq \lambda^* \leq \lambda_{max}$ and $\lambda^* \neq 1$ and $\lambda = \lambda^*$.

Proposition 3.6 implies that, under ***Case 2*** and ***Case 3***, only if there exists an unique perturbation ratio and the target-perturbation branch is perturbed to that value, the dimension of the stealthy attack space $dim(\mathcal{A}_s)$ increases 1. Based on this result, if $\Delta\gamma = 0$ when we perturb a target-perturbation branch to a certain value, then this result almost remains the same regardless of the perturbation magnitude. Further, we obtain a result about the increase of the security factor.

Theorem 3.3 Under all cases, $\Delta\gamma = 1$ for any $\lambda > 0$ *(*$\lambda \neq 1$*)* if and only if there exists $\lambda = \lambda^*$ *(*$\lambda^* \neq 1$*)* such that $\Delta\gamma = 1$.

Proof <u>***Case 1*** : $i_d \notin \mathcal{M}^q_{mtd}$ ***and*** $j_d \notin \mathcal{M}^q_{mtd}$</u> In ***Case 1***, we have proved that the susceptance-perturbation magnitude does not affect the change of security factor γ. Therefore, if there exists $\lambda = \lambda^*$ such that $\Delta\gamma = 1$, then this result will not change if we change the perturbation magnitude to the other values. The proof of necessity is obvious.

<u>***Case 2*** : $i_d \in \mathcal{M}^q_{mtd}$ ***and*** $j_d \notin \mathcal{M}^q_{mtd}$</u> (Sufficiency) Suppose there exists a column $\mathbf{h}^p_t$ in $\mathbf{H}^p$ such that $\mathbf{h}^p_t$ cannot be linearly represented by the columns in $[\mathbf{H}\ \mathbf{H}^q_r]$ when we perturb the susceptance of branch k_d to $\lambda^* b_{i_d j_d}$, where $\mathbf{H}^q_r$ is a submatrix formed by the columns of $\mathbf{H}^q$ that are not changed after the perturbation. Since $R([\mathbf{H}\ \mathbf{H}^q]) = R([\mathbf{H}\ \mathbf{H}'])$, we cannot find another two columns in $\mathbf{H}^p$ that cannot be linearly represented by the columns in $[\mathbf{H}\ \mathbf{H}^q_r]$. Therefore, in order to have $\Delta\gamma = R([\mathbf{H}\ \mathbf{H}'']) - R([\mathbf{H}\ \mathbf{H}']) = 1$, at least one of $\mathbf{h}''_{i_d}$ and $\mathbf{h}''_{j_d}$ cannot be linearly represented by the columns in $[\mathbf{H}\ \mathbf{H}^q_r]$, where $\mathbf{h}''_{i_d}$ and $\mathbf{h}''_{j_d}$ are respective the i_dth column and j_dth column in $\mathbf{H}''$ after the perturbation of branch k_d. Here, we suppose that $\mathbf{h}''_{i_d}$ cannot be linearly represented by the columns in $[\mathbf{H}\ \mathbf{H}^q_r]$ (we can draw the same conclusion if $\mathbf{h}''_{j_d}$ cannot be linearly represented by the columns in $[\mathbf{H}\ \mathbf{H}^q_r]$). Let $\mathbf{Q} = [\mathbf{H}\ \mathbf{H}^q_r\ \mathbf{h}^p_t\ \mathbf{h}''_{i_d}]$ be a matrix of full column rank and $R(\mathbf{Q}) = R([\mathbf{H}\ \mathbf{H}'']) = R([\mathbf{H}\ \mathbf{H}']) + 1$. We select $(n+q+1)$ rows that contain the perturbed rows (i.e., there are elements $\lambda^* b_{i_d j_d}$ and/or $-\lambda^* b_{i_d j_d}$ in this row) and $(n+q+1)$ columns from $\mathbf{Q}$ to form an $(n+q+1)$ by $(n+q+1)$ square matrix $\bar{\mathbf{Q}}$. Using the Leibniz formula, we can derive that $\det(\bar{\mathbf{Q}})$ is a linear function of $\lambda^* b_{i_d j_d}$, that is, $\det(\bar{\mathbf{Q}}) = ab_{i_d j_d}\lambda^* + b$, where a and b are values after the calculation. Specifically, a and b are different if we choose different rows and columns in $\mathbf{Q}$ to form $\bar{\mathbf{Q}}$. Since $\Delta\gamma = 1$, we can derive that there exists $\bar{\mathbf{Q}}$ such that $\det(\bar{\mathbf{Q}}) = ab_{i_d j_d}\lambda^* + b \neq 0$. Therefore, a and b cannot be 0 at the same time.

If $a = 0$, we have $b \neq 0$. Therefore, for any λ^*, we have $\det(\bar{\mathbf{Q}}) = ab_{i_d j_d}\lambda^* + b \neq 0$. It follows that we always have $\Delta\gamma = 1$ for any $\lambda^* > 0$ $(\lambda \neq 1)$. If $a \neq 0$, we analyze whether or not there exists λ such that $\Delta\gamma < 1$. If $\Delta\gamma < 1$, we must have $R(\mathbf{Q}) \leq R([\mathbf{H}\ \mathbf{H}'])$. It indicates that, for any $\bar{\mathbf{Q}}$, we have $\det(\bar{\mathbf{Q}}) = 0$. As we know, $\lambda^* = 1$ is the unique solution for $\det(\bar{\mathbf{Q}}) = ab_{i_d j_d}\lambda^* + b = 0$ (for any $\bar{\mathbf{Q}}$). However, since $\lambda^* \neq 1$, we can never have $\det(\bar{\mathbf{Q}}) = 0$ for any $\bar{\mathbf{Q}}$. Therefore, $\Delta\gamma = 1$ for any $\lambda > 0$ $(\lambda \neq 1)$.

(Necessity) Obviously, if $\Delta\gamma = 1$ for any $\lambda > 0$ $(\lambda \neq 1)$, there must exists $\lambda = \lambda^*$ such that $\Delta\gamma = 1$.

<u>***Case 3*** : $i_d \in \mathcal{M}^q_{mtd}$ ***and*** $j_d \in \mathcal{M}^q_{mtd}$</u> (Sufficiency) Let $\mathbf{P} = [\mathbf{H}\ \mathbf{H}^q_r]$, where $\mathbf{H}^q_r$ is a submatrix containing the columns in $\mathbf{H}^q$ that are not changed after perturbing the susceptance of branch k_d to $\lambda^* b_{i_d j_d}$. Because $R([\mathbf{H}\ \mathbf{H}^q]) = R([\mathbf{H}\ \mathbf{H}']) = n + q$, we can never find another three columns in $\mathbf{H}^p$ to form a combined matrix with $\mathbf{P}$ that has rank $n+q+1$. Therefore, supposing $R([\mathbf{H}\ \mathbf{H}^q_r\ \mathbf{h}''_t\ \mathbf{h}''_{t'}\ \mathbf{h}''_{t''}]) = n+q+1$, then at least one column in $\mathbf{h}''_t, \mathbf{h}''_{t'}, \mathbf{h}''_{t''}$ is either $\mathbf{h}''_{i_d}$ or $\mathbf{h}''_{j_d}$, where $\mathbf{h}''_{i_d}$ and $\mathbf{h}''_{j_d}$ are the columns

being perturbed in $\mathbf{H}^q$ after the perturbation of branch k_d. Here we assume that is $\mathbf{h}''_{i_d}$ (we can draw the same conclusion if it is $\mathbf{h}''_{j_d}$). Let $\mathbf{Q} = [\mathbf{H}\ \mathbf{H}_r^q\ \mathbf{h}''_{i_d}\ \mathbf{h}''_t\ \mathbf{h}''_{t'}]$. It follows that $R(\mathbf{Q}) = n + q + 1$. We select $(n + q + 1)$ rows that contain the perturbed rows (i.e., there are elements $\lambda^* b_{i_d j_d}$ and/or $-\lambda^* b_{i_d j_d}$ in this row) and $(n + q + 1)$ columns from $\mathbf{Q}$ to form an $(n + q + 1)$ by $(n + q + 1)$ square matrix $\bar{\mathbf{Q}}$. Using the Leibniz formula, we can obtain that the determinant of $\bar{\mathbf{Q}}$ is $\det(\bar{\mathbf{Q}}) = ab_{i_d j_d}\lambda^* + b$, where a and b are coefficients after the calculation. Since there exists $\bar{\mathbf{Q}}$ such that $\det(\bar{\mathbf{Q}}) = ab_{i_d j_d}\lambda^* + b \neq 0$, a and b cannot be 0 at the same time.

If $a = 0$, we have $b \neq 0$. Thus, $\det(\bar{\mathbf{Q}}) \neq 0$ for any $\lambda^* > 0$. Therefore, $\Delta\gamma = 1$ for any $\lambda^* > 0$. If $a \neq 0$, as we know, $\lambda^* = 1$ is the unique solution for $\det(\bar{\mathbf{Q}}) = ab_{i_d j_d}\lambda^* + b = 0$. However, since $\lambda^* \neq 1$, we can never have $\det(\bar{\mathbf{Q}}) = ab_{i_d j_d}\lambda^* + b = 0$ for any $\bar{\mathbf{Q}}$ and any $\lambda > 0$ $(\lambda^* \neq 1)$. Therefore, $\Delta\gamma = 1$ for any $\lambda > 0$ $(\lambda \neq 1)$.

(Necessity) Since $\Delta\gamma = 1$ for any $\lambda > 0$ $(\lambda \neq 1)$, we must have $\lambda = \lambda^*$ such that $\Delta\gamma = 1$.

Theorem 3.3 implies that if the security factor γ increases 1 when we perturb one more branch, then this result will not be changed with the variation of the susceptance perturbation magnitude. In other words, once we find that the dimension of the stealthy attack space $dim(\mathcal{A}_s)$ decreases 1 in a trial, then this result will not change in the other trials. Therefore, we can determine the security factor γ using only one tested perturbation ratio when $\Delta\gamma = 1$.

In summary, the susceptance perturbation magnitude almost does not affect the change of the dimension of the stealthy attack space $dim(\mathcal{A}_s)$ when one more branch is perturbed. Thus, we can determine the dimension of the stealthy attack space after we have perturbed the branches to certain ratios, without worrying about the impact of the perturbation magnitude, especially the case when $\Delta\gamma = 1$. Note that this is very useful for the selection of the target-perturbation branches. We present details about this issue in the next subsection.

3.3.4 Guidance on Constructing Effective Moving Target Defense

In this section, we first consider the optimal selection of the set of target-perturbation branches. Then, we discuss the increasing generation cost caused by MTD.

Target-Perturbation Branch Selection

Here we propose an algorithm for maximizing the security factor γ (i.e., minimizing the dimension of the stealthy attack space $\mathcal{A}_s$) and covering the largest number of buses, with a given number of D-FACTS devices. We briefly introduce the algorithm in the following.

Algorithm 1: Minimizing the dimension of the stealthy attack-space

Data: The number of D-FACTS devices n_d; The set of branches that can be perturbed $\mathcal{L}_p$

1 Initialization: randomly rearrange $\mathcal{L}_p$ as $\mathcal{L}_p = \{k'_1, k'_2, \ldots, k'_p\}, r = 0,$ $r' = 0, \mathcal{L}_d^1 = \phi, \mathcal{L}_d^2 = \phi, \mathcal{L}_d^3 = \phi, \mathcal{L}_d = \phi;$

2 **for** *each* $k'_t \in \mathcal{L}_p$ **do**

3 | Perturb k'_t to a random ratio λ;

4 | **if** $r = n_d$ **then**

5 | | break ;

6 | **end**

7 | Compute $\Delta\gamma = R([\mathbf{H}\ \mathbf{H}'']) - R([\mathbf{H}\ \mathbf{H}'])$;

8 | **if** $\Delta\gamma = 1$ **then**

9 | | put k'_t in the branch set $\mathcal{L}_d^1$;

10 | | $r = r + 1$;

11 | **end**

12 **end**

13 **if** $r < n_d$ **then**

14 | Enter Algorithm 2;

15 | $\mathcal{L}_d = \mathcal{L}_d^1 \cup \mathcal{L}_d^2$;

16 **else**

17 | $\mathcal{L}_d = \mathcal{L}_d^1$;

18 **end**

Result: The set of target-perturbation branches $\mathcal{L}_d$

For the sake of power transfer quantity and quality, the branch parameters of some branches cannot be perturbed. Therefore, the set of candidate branches input to Algorithm 1 may not be the set of all branches in the power network. Thus, we only need to seek for an optimal deployment strategy in the set of branches that can be perturbed. With the loop (form line 2 to line 12) in Algorithm 1, we traverse all perturbable branches until we obtain the maximum γ. This process computes a set of target-perturbation branches that can minimize the stealthy attack space. In the loop, we determine whether the branch should be selected or not from line 8 to line 11. Once $\Delta\gamma = 1$, the branch is selected as a target-perturbation branch; otherwise, it is not. When the iterations in Algorithm 1 have finished, we determine whether we should go into Algorithm 2 with the condition given in line 13. That is, if γ is maximized and all the D-FACTS devices have been used, then the algorithm is closed and the output $\mathcal{L}_d^1$ is the set of target-perturbation branches. If γ is maximized but there still exist unused D-FACTS devices, the algorithm enters Algorithm 2 for maximizing the covered buses.

Algorithm 2 starts with the rest candidate branches except those have been selected for maximizing γ. It incrementally searches for branches that cover new buses. The target-perturbation branch is selected according to the following process. First of all, we select the branches that cover two new buses (line 7 to line 10). Then, if there still exist unused D-FACTS devices, we select the branches that cover one new bus (line 16 and line 17). At last, if there still exist unused D-FACTS devices, we select the branches from the rest candidate branches (line 19 to line 21).

The computation time requirement for the rank operation of $\Delta\gamma = R([\mathbf{H}\,\mathbf{H}''])-R([\mathbf{H}\,\mathbf{H}'])$ is $O(mn^2)$, where m is the number of measurements and n is the number of buses in the power network. The computational complexity for Algorithm 2 is $O(l)$, where l is the number of branches in the power network. Considering the worst case that the rank operation would be executed l times in the loop, the time complexity for Algorithm 1 is $O(lmn^2)$. Note that Algorithm 2 is not executed if the condition given in line 13 of Algorithm 1 is not satisfied. In fact, the above algorithm only outputs an alternative set of target-perturbation branches. There may be a lot of candidates that are also satisfied (see Sect. 3.4.3). Therefore, we can dynamically change the perturbed branches and maintain the effectiveness of MTD. On the other hand, we can realize MTD on the basis of already deployed D-FACTS devices for saving the additional infrastructure cost.

Reducing Operation Cost

After determining the set of branches that should be perturbed, a following problem that we should consider is to reduce the operation cost by appropriately setting their perturbation magnitudes. The OPF seeks to optimize the operation of an electric power system subject to the physical constraints imposed by electrical laws and

Algorithm 2: Covering the largest number of buses

Compute: $\mathcal{L}'_p = \mathcal{L}_p \setminus \mathcal{L}^1_d$; the set of buses $\mathcal{I}$ covered by branches in $\mathcal{L}^1_d$;
for *each* $k''_t = \{i''_t, j''_t\} \in \mathcal{L}'_p$ **do**
    if $r + r' = n_d$ **then**
        break ;
    end
    if i''_t *or* $j''_t \notin \mathcal{I}$ **then**
        if i''_t *and* $j''_t \notin \mathcal{I}$ **then**
            put k''_t in the branch set $\mathcal{L}^2_d$;
            $r' = r' + 1$;
        else
            put k''_t in the branch set $\mathcal{L}^3_d$;
        end
    end
end
if $r + r' < n_d$ **then**
    if $n_d - r - r' \leq |\mathcal{L}^3_d|$ **then**
        Select $n_d - r - r'$ branches from $\mathcal{L}^3_d$ and put them into the branch set $\mathcal{L}^2_d$;
    else
        Put all branches in $\mathcal{L}^3_d$ into $\mathcal{L}^2_d$;
        Select $n_d - r - r' - |\mathcal{L}^3_d|$ branches from $\mathcal{L}'_p \setminus \mathcal{L}^2_d$ and put them into $\mathcal{L}^2_d$;
    end
end
Return $\mathcal{L}^2_d$;

engineering limits [12]. It outputs the minimum generation cost for the given loads by adjusting the power flows. This generation cost can represent the operation cost caused by the system changes of MTD. Here, OPF is stated as follows:

$$
\begin{aligned}
\min_{p_i^g, i\in\mathcal{G}} \quad & \sum_{i\in\mathcal{G}} C_i(p_i^g) \\
s.t. \quad & p_i^g - p_i^l = \mathbf{B}\boldsymbol{\theta},\ i\in\mathcal{N} \\
& p_i^{min} \le p_i^g \le p_i^{max},\ i\in\mathcal{G} \\
& f_{ij}^{min} \le f_{ij} \le f_{ij}^{max},\ \{i,j\}\in\mathcal{L} \\
& \lambda_{ij}^{min} \le \lambda_{ij} \le \lambda_{ij}^{max},\ \{i,j\}\in\mathcal{L}
\end{aligned}
\tag{3.23}
$$

where $C_i(p_i^g)$ is the cost function, $\mathcal{G}$ is the set of generators, p_i^g is the real output power, p_i^l is the load, f_{ij} is the branch active power flow, λ_{ij} is the susceptance perturbation ratio. In the above optimization problem, the decision variable is the output generation level of each generator, and the cost function is a quadratic function. The first constraint is about the nodal power balance constraint, i.e., the power generations must be equal to the power consumptions. The second and third constraints are about the limits about the output generation and branch active power flow, respectively. The fourth constraint is newly added here for the limits of the susceptance perturbation. We can see that the matrix **B** contained in the first constraint contains the branch perturbation parameters. Therefore, this optimization problem is correlated with the branch perturbation magnitude. Since the objective function is convex and the constraints are differentiable, the OPF problem is a typically convex optimization problem [3, 47], which can be solved by the non-linear programming solver *fmincon* in MATLAB.

3.4 Simulation Results

In this section, we evaluate our findings about MTD using an illustrative IEEE 14-bus power system (Fig. 3.5) and the IEEE 30-bus, 57-bus, 118-bus and 145-bus power systems. All simulations are based on the fully measured power system and carried out in MATLAB. We assume that all meters are subject to the same noise distribution, namely the Normal distribution $\mathcal{N}(0, \sigma^2)$, if the measurement noises are considered. The threshold τ (see Sect. 3.1) is set as $\sigma\sqrt{\chi^2_{m-n,\alpha}}$ [47], where $m-n$ is the freedom degree of the Chi-square distribution and α is the false alarm rate (which is 0.05). In practice, each branch susceptance must be within given limits, namely $b_{ij}^{min} \le b_{ij} \le b_{ij}^{max}$ [45]. The authors in [34] have proved that if the perturbations are within 20% of the impedance, then there are sufficiently large number of perturbation cases that can restrict power losses within 1%. Thus, it is feasible to perturb the branch

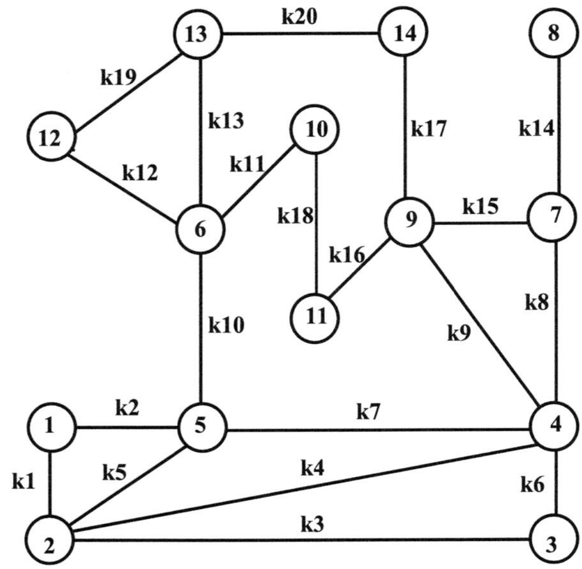

Fig. 3.5 The IEEE 14-bus power system

susceptances within 20% maximum change. In this chapter, for the perturbation ratio λ, it is constrained to be within [0.8, 1.2].

3.4.1 Effectiveness of Moving Target Defense

Dimension of Stealthy Attack Space Versus Branch Perturbations

Taking the IEEE 14-bus power system (Fig. 3.5) as an example, we analyze the dimension change of the stealthy attack space ($\mathcal{A}_s$) when we increase the number of perturbed branches. We successively perturb the set of branches as $\mathcal{L}_1^d = \{k_1\}$, $\mathcal{L}_2^d = \{k_1, k_2\}$, $\mathcal{L}_3^d = \{k_1, k_2, k_3\}, \ldots$. Figure 3.6 shows the simulation results. We can see that the dimension of the stealthy attack space decreases when more branches are perturbed. But $dim(\mathcal{A}_s)$ cannot reach 0, because the 14-bus power system only has 20 branches, which is less than $2n = 26$. Even though we perturb all branches, the dimension of the stealthy attack space is 6, which is the smallest stealthy attack space we can achieve in this 14-bus power system. Consistent with Eq. (3.14), the smallest dimension of the stealthy attack space is equal to $2n - l = 2 * 13 - 20 = 6$.

Moreover, we perturb all branches in the IEEE 30-bus, 57-bus, 118-bus and 145-bus power systems. Table 3.2 shows the smallest dimension of the stealthy attack space we obtain. We also give the number of branches for each power system in this table. We can see that all the smallest dimensions of the stealthy attack space are not zero, which indicates that none of the power systems support constructing complete MTD. Even though the IEEE 145-bus power system has 453 branches,

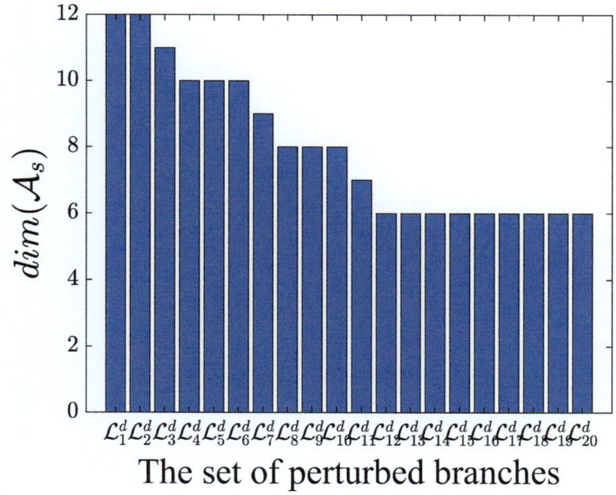

Fig. 3.6 The dimension change of the stealthy attack space $\mathcal{A}_s$ with respect to the perturbed branches

Table 3.2 The dimension of $\mathcal{A}_s$ after perturbing all branches

IEEE test system	30-bus	57-bus	118-bus	145-bus
# branches	41	80	186	453
$dim(\mathcal{A}_s)$	18	31	40	10

which is larger than $2n = 288$, the dimension of the stealthy attack space is 10 after perturbing all its branches. In our opinion, one reason for this result is that the IEEE 145-bus power system contains 7 buses that are only connected by a single branch. The simulation result indicates that is difficult to meet the conditions required for achieving a complete MTD in practice.

Effectiveness of Moving Target Defense for Thwarting False Data Injection Attacks

Next, we exploit MTD's effectiveness for thwarting FDI attacks with respect to the dimension of the stealthy attack space. The attack vector is constructed with the form of $\mathbf{a} = \mathbf{Hc}$, where $\mathbf{H}$ is a measurement matrix before MTD. We sample the value of the element c_i in $\mathbf{c}$ from a uniform distribution $\mathcal{U}(-dm, dm)$, where dm is the maximum magnitude of the injected bias into the state variable. Here dm is 0.1. The modified state variables are uniformly selected from the bus set, i.e., the non-zero elements in $\mathbf{c}$ are uniformly selected. We assume that all branches in the power system can be perturbed, and the measurements are noiseless. The perturbation ratio is randomly chosen within [0.8, 1.2]. For each setting, we repeat random attacks for

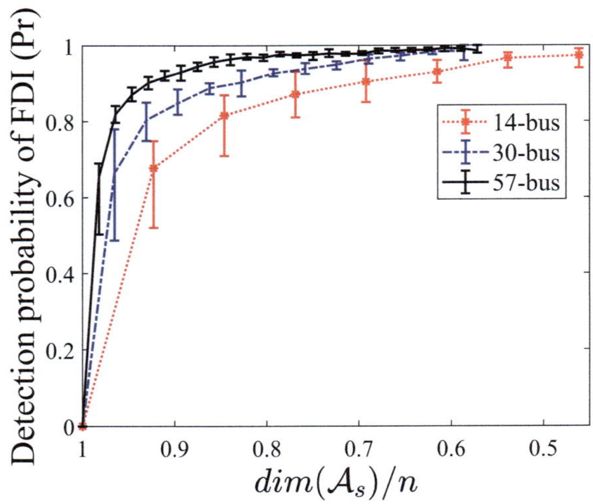

Fig. 3.7 The detection probability of FDI attacks with respect to the dimension of the stealthy attack space

1000 times based on Monte Carlo simulations, and estimate the detection probability of the FDI attack constructed with **H** as

$$\text{Pr} = \frac{\text{\# of detected trials}}{1000}, \tag{3.24}$$

where *# of detected trials* means the number of FDI attack vectors detected by the BDD after MTD. For consistency, we use the metric $\frac{dim(\mathcal{A}_s)}{n}$ to measure the change of the dimension of the stealthy attack space. Note that for a fixed dimension of the stealthy attack space, there are several branch-perturbation schemes for realizing MTD.

We use the IEEE 14-bus, 30-bus and 57-bus power systems with default settings and data in MATPOWER. Figure 3.7 shows the simulation results. We can see that if the dimension of the stealthy attack space is smaller, the detection probability of FDI attacks is larger. For example, in the 14-bus power system, the detection probability of FDI is more than 90% when $\frac{dim(\mathcal{A}_s)}{n}$ is reduced to 0.6. Moreover, we find that if the system size is larger, the curve is more smooth and the detection probability of FDI attacks is larger given the same $\frac{dim(\mathcal{A}_s)}{n}$. The simulation results highlight that the smaller the dimension of the stealthy attack space is, the better performance the MTD achieves in terms of thwarting FDI attacks.

Besides, we consider the case when the attacker has limited resources to modify state variables. We take the IEEE 14-bus power system as an example. First, we assume that the attacker can only modify 3 state variables, that is, there are only 3 non-zero elements in **c**. We analyze the change of the detection probability of FDI attacks with respect to the dimension of the stealthy attack space. For each setting, we repeat random attacks for 1000 times based on Monte Carlo simulations. That is, for each time, the attacked state variables and the attack vector are randomly generated. Figure 3.8 shows the simulation result. We can see that the detection probability

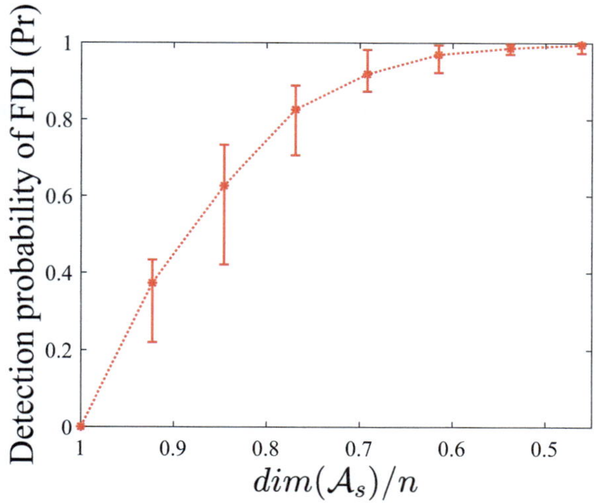

Fig. 3.8 The detection probability of FDI attacks when only 3 state variables can be modified by the attacker

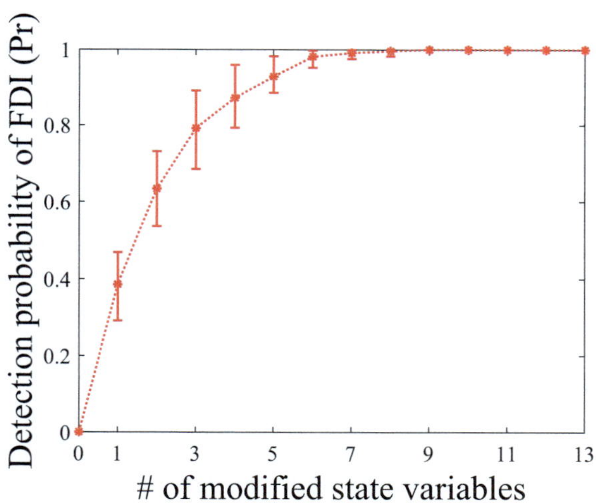

Fig. 3.9 The detection probability of FDI attacks when the dimension of the stealthy attack space is fixed

increases with $\frac{dim(\mathcal{A}_s)}{n}$ decreases. Again, it proves that a smaller stealthy attacks space is more effective to thwart FDI attacks. Second, we assume that the dimension of the stealthy attack space is fixed as 10, and the deployment of D-FACTS devices is determined, while the number of modified state variables is varied. Note that there are several combinations for a specific number of state variables. For each setting, we repeat the simulation for 1000 times and uniformly select a set of state variables for each time. The simulation result is plotted in Fig. 3.9. We can see that the detection probability of FDI attacks increases with the number of modified state variables. This indicates that the attacker is more ambitious, the MTD is more effective.

Moreover, we investigate the impact of FDI attacks constructed with the former measurement matrix on the buses that are not covered by the perturbed branches.

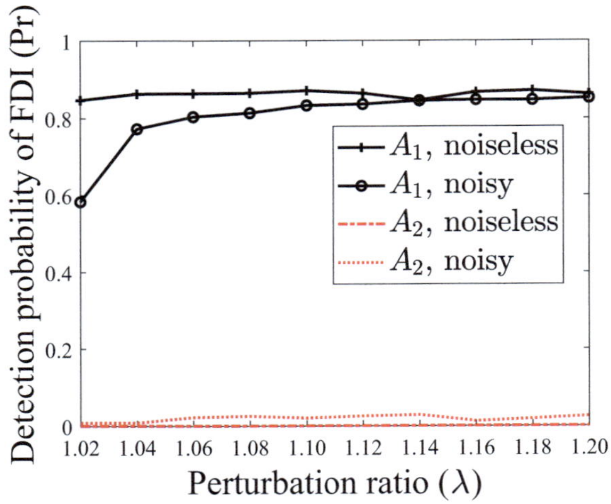

Fig. 3.10 The detection probability of FDI attacks with $c_i \neq 0$ for $i \in \mathcal{M}_{mtd}$ (A_1) and $c_i = 0$ for all $i \in \mathcal{M}_{mtd}$ (A_2)

We select k_1, k_3, k_4 and k_7 as the target-perturbation branches. They are perturbed to the same ratio during simulations. The perturbation ratios are set within [1.02, 1.2] spaced by 0.02. The attack vector $\mathbf{a} = \mathbf{Hc}$ is constructed by considering two cases: $c_i \neq 0$ for $i \in \mathcal{M}_{mtd}$ and $c_i = 0$ for all $i \in \mathcal{M}_{mtd}$, where $\mathcal{M}_{mtd}$ is a set of buses that are covered by the perturbed branches. We denote these two cases as "A_1" and "A_2," respectively. The measurement noise is fixed as $\sigma^2 = 0.01$. For each setting, we repeat the simulation for 1000 times. For each time, we uniformly selected the non-zero elements in $\mathbf{c}$. The simulation results are given in Fig. 3.10. We can see that, for case A_1, the detection probability of the FDI attack is more than 84% when the measurements are noiseless. We believe that this large detection probabilities is because $\mathbf{a} \notin S(\mathbf{H}')$ holds for most cases. When the measurements are noisy in case A_1, the detection probability of FDI attacks is disturbed by the noise. But it is still more than 60% and increases with the perturbation ratio. For case A_2, we can see that, whether the measurements are noisy or not, the detection probability of FDI attacks is almost negligible. This indicates that the buses that are not covered by the perturbed branches are vulnerable to FDI attacks. Therefore, to improve the effectiveness of MTD, the perturbed branches should cover as many buses as possible.

Impact of Bus Connected by Single Branch

Next, we consider the attack on the bus that is only connected by a single branch. Here we validate our finding using the IEEE 14-bus power system. We find that bus 8 is only connected by branch {7, 8}. First, we only inject errors into the state variable of bus 8. We test this attack for 10 trials. For each trial, we select 1000 sets of perturbed branches containing branch {7, 8} randomly. The noise variance is fixed as 0.01. Figure 3.11 shows the simulation results. We can see that the attacks on bus 8 can always bypass the BDD after MTD when the measurements are noiseless. And for

Fig. 3.11 The detection probability of FDI attacks versus different sets of perturbed branches with only bus 8 is attacked

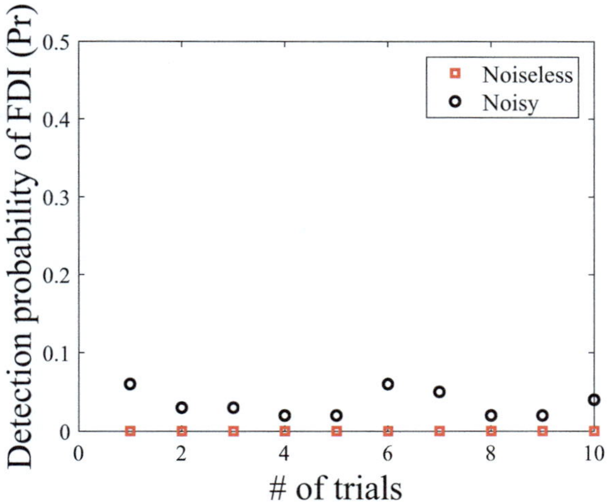

Fig. 3.12 The attacks on bus 8 can always bypass the BDD after MTD in the IEEE 14-bus power system

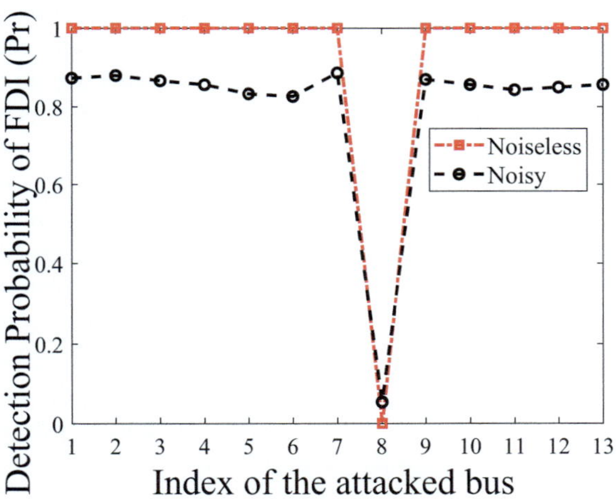

the noisy case, the detection probability of this attack is almost negligible (around 0.05). By contrast, we conduct another 13 trials but focus on attacking different buses. For the ith trial, we only modify the state variable of bus i. In these trials, we perturb all branch susceptances. Figure 3.12 shows the simulation results. We can see that, no matter the measurements are noiseless or noisy, we have a negligible detection probability of the FDI attack when the state variable of bus 8 is modified. But it can be detected when we attack the other buses. The above results highlight the weakness of the bus that is only connected by a single branch and prove that we can never achieve a complete MTD when the power network contains such a bus.

Next, we consider the induced generation cost when attacking the bus that is only connected by a single branch. Considering the IEEE 30-bus power system, there

Table 3.3 Parameters of the generators

Generation bus	1	2	13	22	23	27
P^g_{max}	100	100	100	100	100	100
μ_i($/MWh)	20	30	20	20	30	20

Table 3.4 Increasing in generation cost when the load bus that is only connected by a single branch is attacked

Communicated load	L_0	$1.2L_0$	$1.4L_0$	$1.6L_0$	$1.8L_0$	$2.0L_0$
Generation cost $(10^3\$/h)$	3.784	3.798	3.812	3.826	3.840	3.854
Increasing rate	0	0.37%	0.74%	1.11%	1.48%	1.85%

are 3 buses that are only connected by a single branch. And one of them is a load bus (bus 26). In the following, we exploit the increasing generation cost when the attacker compromises this load bus. The objective function is linear with the value of generation, i.e., $C_i(p_i^g) = \mu_i p_i^g$ [see the OPF problem (3.23)]. The parameters about the generators are shown in Table 3.3. And all active power flows are limited to 500 MW. In this case, we perturb all branches with the perturbation ratios sampled within [0.8, 1.2]. The default setting and load data provided in MATPOWER are used. First, we use *fmincon* in MATLAB to compute an optimal generation dispatch result. Then, the measurements associated with bus 26 are corrupted. It results in deceiving the amount of the load on bus 26 transmitted to the control center for solving the OPF problem. The simulation results are shown in Table 3.4. L_0 is the original load at bus 26. We can see the generation cost increases with the corrupted load. Actually, the load attack may also lead the system to a non-optimal generation dispatch, and the worst, may cause load shedding [28].

3.4.2 *Impact of Susceptance Perturbation Magnitude on Dimension of Stealthy Attack Space*

To demonstrate the results given in Sect. 3.3.3, we analyze the dimension change of the stealthy attack space by varying the perturbation ratio. We give six examples for the three cases stated in Sect. 3.3.3 with the IEEE 14-bus power system. For each example, we randomly choose 10000 perturbation ratios within [0.8, 1.2]. At the beginning, we choose an initial set of perturbed branches. Then, we perturb one more branch k_d. Let $dim(\mathcal{A}_s)$ and $dim^*(\mathcal{A}_s)$ denote the dimension of the stealthy attack space before and after this perturbation, respectively. During simulations, we count the number of exceptions when $dim^*(\mathcal{A}_s)$ is changed. The simulation results are shown in Table 3.5. We can see that if the dimension of the stealthy attack space

Table 3.5 Impact of the perturbation magnitude on the dimension of the stealthy attack space

Cases	Initial set	k_d	$dim^*(\mathcal{A}_s) - dim(\mathcal{A}_s)$	# Exceptions
Case 1	$\{k_1, k_2, k_3, k_4\}$	k_{14}	0	0
	$\{k_1, k_2, k_3, k_4\}$	k_{16}	−1	0
Case 2	$\{k_1, k_2, k_3, k_4\}$	k_5	0	0
	$\{k_5, k_6, k_8\}$	k_1	−1	0
Case 3	$\{k_1, k_2, k_3, k_4\}$	k_6	0	1
	$\{k_1, k_6\}$	k_4	−1	0

is reduced by 1, then this result always holds for all perturbation ratios in these three cases. If the dimension of the stealthy attack space remains the same after the perturbation of branch k_d, we do not find exceptions in the first (***case 1***) and third (***case 2***) example, i.e., this result always holds regardless of the perturbation ratio. But in the fifth (***case 3***) example, we find the dimension of the stealthy attack space increases 1 when we set the perturbation ratio to be 0.88. In other words, $dim(\mathcal{A}_s) = dim^*(\mathcal{A}_s)$ does not always hold in ***case 3***. Moreover, we find that this exception happens when the perturbation ratio of branch k_6 is the same with that of branch k_3. Therefore, in order to avoid the change of stealthy attack space's dimension with the perturbation magnitude, we should not perturb the branches to the same ratio.

3.4.3 Guidance on Constructing Effective Moving Target Defense

Target-Perturbation Branch Selection

Evaluation of algorithm. First, with the IEEE 14-bus power system, we evaluate the selection of the set of target-perturbation branches with Algorithms 1 and 2. A total number of 10 D-FACTS devices are given for constructing/realizing MTD. We assume that all branches in this power system can be perturbed. With the aim to minimize the dimension of the stealthy attack space and maximize the number of covered buses, we adopt Algorithms 1 and 2 in Section IV-D1. From Algorithm 1, we obtain a set of target-perturbation branches as $\{k_1, k_3, k_4, k_7, k_8, k_{11}, k_{12}\}$. From Algorithm 2, we obtain a set of target-perturbation branches as $\{k_{16}, k_{17}, k_{20}\}$. We get the final set of the target-perturbation branches by combining these two sets. With this MTD, the dimension of the stealthy attack space is 6 and the number of covered buses is 13. The target-perturbation branches deployed with D-FACTS devices are shown in Fig. 3.13a. For comparison, if we optionally perturb the set of branches $\{k_1, k_2, k_3, k_4, k_5, k_6, k_7, k_8, k_9, k_{10}\}$ (Fig. 3.13b), the dimension of stealthy attack space is 8 and the number of the covered buses is 8. This indicates that the output from our algorithm is better than the optional selections.

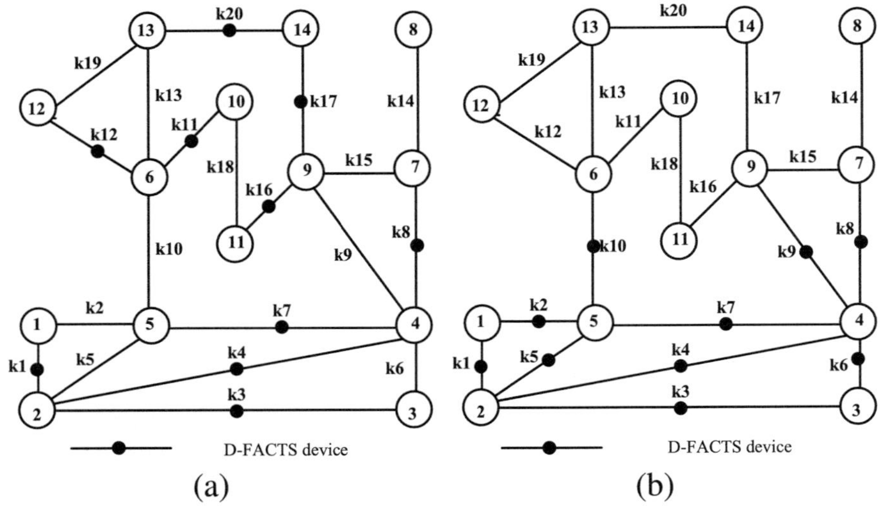

Fig. 3.13 Comparison of the optimal deployment of D-FACTS devices using our algorithm and the optionally selected target-perturbation branches. **a** Output from the Algorithms given in Sect. 3.3.4; **b** Optionally selected target-perturbation branches. In the first case, $\dim(\mathcal{A}_s) = 6$ and the number of covered buses is 13; while in the second case, $\dim(\mathcal{A}_s) = 8$ and the number of covered buses is 8

In fact, if we start Algorithm 1 from different branches, we can obtain different sets of target-perturbation branches. We take the IEEE 14-bus power system as an example. Suppose that there are 9 D-FACTS devices for constructing MTD, and all branches in this 14-bus power system can be perturbed. We start Algorithm 1 from branch k_2 and k_{16}, respectively. Correspondingly, we obtain the sets of target-perturbation branches as $\mathcal{L}_d^{\text{from } k_2} = \{k_2, k_3, k_7, k_9, k_{10}, k_{11}, k_{15}, k_{16}, k_{19}\}$ and $\mathcal{L}_d^{\text{from } k_{16}} = \{k_2, k_4, k_6, k_8, k_{10}, k_{11}, k_{16}, k_{19}, k_{20}\}$, respectively. We can see that the sets of target-perturbation branches are different when we start Algorithm 1 from branch k_2 and k_{16}. But we find that the values of γ and the number of covered buses are the same, i.e., $\gamma = 7$ and the number of covered buses is 13. This indicates that, even though the sets of target-perturbation branches output from our algorithm are different with different start points, they result in the same dimension of stealthy attack space and the same number of covered buses. Thus, we can dynamically change the perturbed branches and maintain the effectiveness of MTD. On the other hand, we can select the branches that are already deployed with D-FACTS devices for realizing MTD, which can help reducing the infrastructure cost.

Computation time of algorithm. Moreover, we evaluate the computation time of the algorithm for the selection of target-perturbation branches. Note that Algorithm 2 is not executed if the condition given in line 13 of Algorithm 1 is not satisfied. We run the algorithm in a core i7 laptop, which has a 2.4GHz CPU and 8.0G memory. We assume that all branches can be perturbed in the adopted power systems. First, we use the IEEE 14-bus power system as an example. We vary the number of perturbed

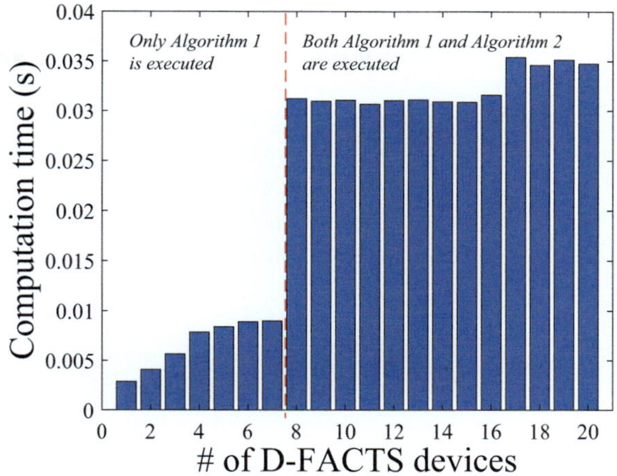

Fig. 3.14 The variation of the computation time of the proposed algorithm with different numbers of D-FACTS devices (14-bus power system)

branches from 1 to 20. The computation time of our algorithm is shown in Fig. 3.14. We can see that the computation time is less than 10ms when the number of D-FACTS devices is less than 8, while it is more than 30ms when the number of D-FACTS devices is larger than 8. It seems that the computation time increases 3 times when the number of D-FACTS devices increases from 8 to 9. This is because only Algorithm 1 is executed when the number of D-FACTS devices is less than 8, i.e., the condition in line 13 of Algorithm 1 is not satisfied. While both Algorithms 1 and 2 are executed when the number of D-FACTS devices is larger than 8. This indicates that sometimes it takes longer for executing Algorithm 2.

Second, we change the size of the power system while fix the number of D-FACTS devices as 10. Here we adopt the IEEE 14-bus, 30-bus, 57-bus, 118-bus and 145-bus power systems. The computation time of our algorithm is given in Fig. 3.15. We

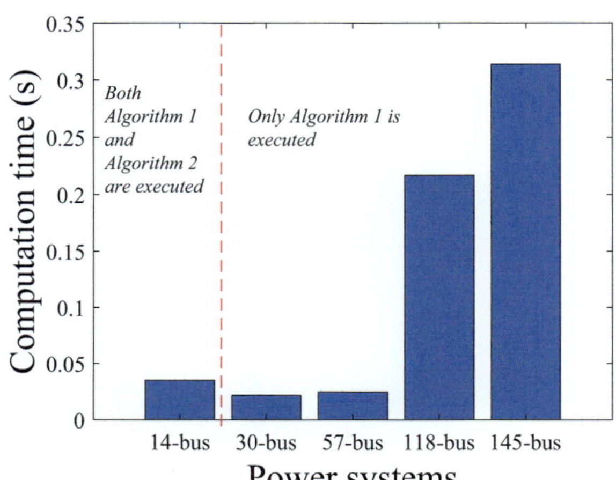

Fig. 3.15 The variation of the computation time of the proposed algorithm with different power systems (given 10 D-FACTS devices)

can see that the computation time increases from around 30ms (14-bus) to more than 300ms (145-bus), which indicates that the computation time of our algorithm increases with the system size. Moreover, we find that the computation time with the 14-bus power system is a little larger than that with the 30-bus power system. In our opinion, the reason is that, given 10 D-FACTS devices, both Algorithms 1 and 2 are executed with the 14-bus power system, while only the Algorithm 1 is executed with the 30-bus, 57-bus, 118-bus and 145-bus power systems. Overall, we can efficiently compute the result (within 350ms in all given power systems).

Reducing Operation Cost

Furthermore, we evaluate the impact of the perturbed branch and the susceptance perturbation magnitude on the increasing of the operation cost. For the OPF problem, we use the objective function as $C_i(p_i^g) = \mu_i p_i^g$, which is a linear generation cost model. With the IEEE 14-bus power system, the generators are installed at bus 1, 2, 3, 6 and 8, and their parameters are shown in Table 3.6. The active power flow limits of branch 1 is 160 MW and the other active power flows are limited to 60 MW. We assume that the optimal results are obtained at the beginning. Then, we analyze the increase of the generation cost when we perturb branch k_2 and k_4 to different ratios, respectively. The simulation results are plotted in Fig. 3.16. We can see that

Table 3.6 Parameters of the generators

Generation bus	1	2	3	6	8
P_{max}^g	300	50	30	50	20
μ_i($/MWh)	20	30	40	50	35

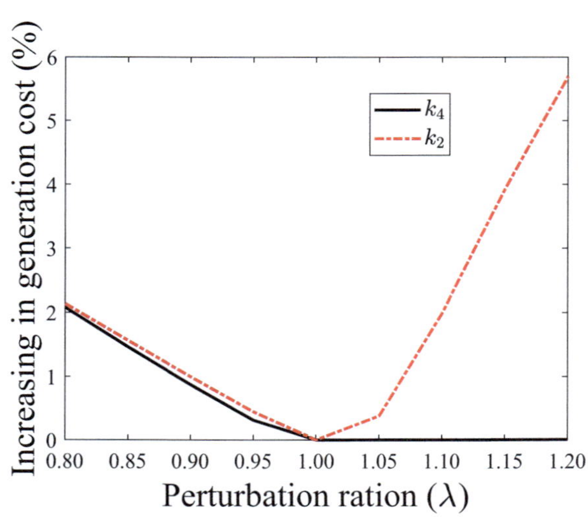

Fig. 3.16 The increasing in generation cost by perturbing branch k_2 and k_4 to different ratios

the increasing of the generation cost under these two cases are different. When the branch susceptance is decreased, the increasing of generation cost by perturbing k_4 is slightly lower than that of by perturbing k_2. But the generation cost increases as the perturbation ratio decreases in both cases. When the branch susceptance is increased, the generation cost of perturbing k_4 almost remains invariant, while it increases with the perturbation ratio by perturbing k_2. The simulation result indicates that, by appropriately selecting the perturbed branch and the perturbation magnitude, we can reduce the increasing operation cost of MTD.

3.5 Conclusion

In this chapter, with the DC power flow model, we analyzed the completeness, deployment and the increasing operation cost of MTD in terms of thwarting stealthy FDI attacks constructed with old system information. To begin with, we proved that an MTD is complete to defeat all FDI attacks constructed with former branch parameters only if the number of branches l is larger than or equal to twice that of the system states n (i.e., $l \geq 2n$, where $n + 1$ is the number of system buses), and the susceptances of more than n branches, which cover all buses, are perturbed. Besides, we proved that we can never realize a complete MTD if the power transmission system has a bus that is only connected by a single branch. Further, we proved that the susceptance perturbation magnitude almost does not affect the dimension of the stealthy attack space after MTD. Based on this result, we presented guidance on effective MTD for minimizing the dimension of the stealthy stealthy attack space, maximizing the number of covered buese and reducing the operation cost. Finally, we illustrated and demonstrated our findings with the IEEE standard test power systems.

References

1. A. Abur, A.G. Expósito, *Power System State Estimation: Theory and Implementation* (Marcel Dekker, Inc., New York, 2004)
2. M.L. Awad, K.E. Casey, A.S. Geevarghese et al., Economic assessment of transmission upgrades: application of the California ISO approach, in *Structured Electric Power System* (Wiley, Hoboken, NJ, USA, 2010), pp. 241–270
3. S. Boyd, L. Vandenberghe, *Convex Optimization* (Cambridge University Press, United Kingdom, 2004)
4. A. Brissette, Performance analysis and design of distributed static series compensators for transmission line susceptance control. Electrical, Computer & Energy Engineering These & Dissertations, University of Colorado, Boulder, CU Scholar, 2014
5. Y. Chen, S. Huang, F. Liu, Z. Wang, X. Sun, Evaluation of reinforcement learning-based false data injection attack to automatic voltage control. IEEE Trans. Smart Grid **10**(2), 2158–2169 (2019)
6. W.-L. Chin, C.-H. Lee, T. Jiang, Blind false data attacks against AC state estimation based on geometric approach in smart grid communications. IEEE Trans. Smart Grid **9**(6), 6298–6306 (2018)

7. K.R. Davis, K.L. Morrow, R. Bobba, E. Heine, Power flow cyber attacks and perturbation-based defense, in *Proceedings of the IEEE International Conference on Smart Grid Communications (SmartGridComm)*, pp. 342–347 (2012)
8. R. Deng, H. Liang, False data injection attacks with limited susceptance information and new countermeasures in smart grid. IEEE Trans. Ind. Inform. **15**(3), 1619–1628 (2019)
9. R. Deng, P. Zhuang, H. Liang, CCPA: coordinated cyber-physical attacks and countermeasures in smart grid. IEEE Trans. Smart Grid **8**(5), 2420–2430 (2017)
10. R. Deng, G. Xiao, R. Lu, H. Liang, A.V. Vasilakos, False data injection on state estimation in power systems-attacks, impacts, and defense: a survey. IEEE Trans. Ind. Inform. **13**(2), 411–423 (2017)
11. D. Divan, H. Johal, Distributed FACTs: a new concept for realizing grid power flow control. IEEE Trans. Power Electron. **22**(6), 2253–2260 (2007)
12. S. Frank, S. Rebennack, An introduction to optimal power flow: theory, formulation, and examples. IIE Trans. **48**(12), 1172–1197 (2016)
13. GE Energy, MAPS^{TM} software-for informed economic decisions. www.gepower.com/prod_serv/products/utility_software/en/downloads/10320.pdf
14. M. Giannini, Improving cyber-security of power system state estimators. M.S. thesis, School Elct. Eng., KTH, Stockholm, Sweden, 2014
15. A. Hamidi, S. Golshannavaz, D. Nazarpour, D-FACTS cooperation in renewable integrated microgrids: a linear multi-objective approach. IEEE Trans. Sustain. Energy 1–9 (2017)
16. A.G. Illera, J.V. Vidal, Lights off! the darkness of the smart meters (2014), http://youtube.be/Z_y_vjYtAWM
17. H. Johal, D. Divan, Design considerations for series-connected distributed FACTS converters. IEEE Trans. Ind. Appl. **43**(6), 1609–1618 (2018)
18. T.T. Kim, H.V. Poor, Strategic protection against data injection attacks on power grid. IEEE Trans. Smart Grid **2**(2), 326–333 (2011)
19. J. Kim, L. Tong, R.J. Thomas, Subspace methods for data attack on state estimation: a data driven approach. IEEE Trans. Signal Process. **63**(5), 1102–1114 (2015)
20. K. Kuntz, M. Smith, K. Wedeward, M. Collins, Detecting, locating, & quantifying false data injections utilizing grid topology through optimized D-FACTS device placement, in *IEEE North American Power Symposium*, pp. 1–6 (2014)
21. M.N. Kurt, Y. Yilmaz, X. Wang, Distributed quickest detection of cyber-attacks in smart grid. IEEE Trans. Inf. Forensics Secur. **13**(8), 2015–2030 (2018)
22. M.N. Kurt, Y. Yilmaz, X. Wang, Real-time detection of hybrid and stealthy cyber-attacks in smart grid. IEEE Trans. Inf. Forensics Secur. **14**(2), 498–513 (2018)
23. S. Lakshminarayana, F. Wen, D.K.Y. Yau, Trade-offs in data-driven false data injection attacks against the power grid, in *IEEE International Conference on Acoustics, Speech and Signal Processing (ICASSP)*, pp. 1–5 (2018)
24. S. Lakshminaraytana, D.K. Yau, Cost-benefit analysis of moving-target defense in power grids (2018). arXiv: 1804.01472
25. S.-C. Lee, J.-Y. Kim, J.-H. Lee, J.-W. Lim, S.-I. Moon, Hybrid linearization of a power system with FACTS devices for a small signal stability study, in *IEEE Power Engineering Society Summer Meeting*, pp. 1566–1572 (2009)
26. F. Li, Y. Wei, S. Adhikari, Improving an unjustified common practice in EX post LMP calculation: an expanded version, in *Proceedings of the IEEE Power Energy Society General Meeting (PESGM)*, pp. 1-4 (2010)
27. B. Li, G. Xiao, R. Lu, R. Deng, H. Bao, On feasibility and limitations of detecting false data injection attack on power grid state estimation using D-FATCS devices. IEEE Trans. Ind. Informat. (2019)
28. J. Liang, L. Sankar, O. Kosut, Vulnerability analysis and consequences of false data injection attack on power system state estimation. IEEE Trans. Power Syst. **31**(5), 3864–3872 (2016)
29. Y. Liu, P. Ning, M.K. Reiter, False data injection attacks against state estimation in electric power grid. ACM Trans. Inf. Syst. Secur. **14**(1), Article no. 13 (2011)

30. B. Liu, H. Wu, A. Pahwa, F. Ding, E. Ibrahim, T. Liu, Hidden moving target defense against false data injection in distribution network reconfiguration, in *Proceedings of the IEEE Power & Energy Society General Meeting (PESGM)*, pp. 1–5 (2018)
31. C. Liu, J. Wu, C. Long, D. Kundur, Reactance perturbation for detecting and identifying FDI attacks in power system state estimation. IEEE J. Sel. Top. Signal Process. **12**(4), 763–776 (2018)
32. D. Mehta, A. Ravindran, B. Joshi, S. Kamalasadan, Graph theory based online optimal power flow control of power grid with distributed flexible AC transmission systems (D-FACTS) Devices, *in IEEE North American Power Symposium*, pp. 1–6 (2015)
33. J. Meng, Z. Yang, Z. Zhang, Y. Geng, R. Deng, P. Cheng, J. Chen, J. Zhou, SePanner: analyzing semantics of controller variables in industrial control systems based on network traffic (distinguished paper award), in *Proceedings of the 39th Annual Computer Security Applications Conference*, pp. 310–323 (2023)
34. K.L. Morrow, E. Heine, K.M. Rogers, R.B. Bobba, T.J. Overbye, Topology perturbation for detecting malicious data injection, in *Proceedings of the International Conference on System Sciences*, pp. 2104–2113 (2012)
35. NERC, Learned lessons (2019). https://www.nerc.com/pa/rrm/ea/Pages/Lessons-Learned.aspx
36. NESCOR, Electric sector failure scenarios and impact analyses version 3.0 (2019). http://smartgrid.epri.com/doc/NESCOR%20Failure%20Scenarios%20v3%2012-11-15.pdf
37. S. Pal, B. Sikdar, J.H. Chow, Classification and detection of PMU data manipulation attacks using transmission line parameters. IEEE Trans. Smart Grid **9**(5), 5057–5066 (2018)
38. PLEXOS software, *Plexos Overview & Tutorial* (2019). www.plexos.info
39. K. Rogers, T. Overbye, Some applications of distributed flexible ac transmission system (D-FACTS) devices in power systems, in *IEEE 40th North American Power Symposium (NAPS)*, pp. 1–8 (2009)
40. S. Sridhar, A. Hahn, M. Govindarasu, Cyber-physical system security for the electric power grid. Proc. IEEE **100**(1), 210–224 (2012)
41. M. Stephen, P. Dmitry, M. Sergei, D. Adam, M. Patrick, Multi-vendor penetration testing in the advanced metering infrastructure, in *Proceedings of the 26th Annual Computer Security Applications Conference*, pp. 107–116 (2010)
42. B. Stott, J. Jardim, O. Alsac, DC power flow revisited. IEEE Trans. Power Syst. **24**(3), 1290–1300 (2009)
43. R. Tan et al., Optimal false data injection attack against automatic generation control in power grid, in *2016 ACM/IEEE 7th International Conference on Cyber-Physical Systems (ICCPS)*, pp. 1–10 (2016)
44. R. Tan et al., Modeling and mitigating impact of false data injection attacks on automatic generation control. IEEE Trans. Inform. Forensics Secur. **12**(7), 1609–1623 (2017)
45. J. Tian, R. Tan, X. Guan, T. Liu, Enhanced hidden moving target defense in smart grids. IEEE Trans. Smart Grids 1–1 (2018)
46. J. Urquizo, P. Singh, N. Kondrath, R. Hidalgo-Leon, G. Soriano, Using D-FACTS in microgrids for power quality improvement: a review, in *IEEE Educator Technical Chapters Meeting (ETCM)*, pp. 1–6 (2017)
47. A. Wood, B. Wollenberg, G.B. Sheblé, *Power Generation, Operation, and Control*, 3rd edn. (John Wiley and Sons, 2013)
48. Q. Yang, D. An, R. Min, W. Yu, X. Yang, W. Zhao, On optimal PMU placement-based defense against data integrity attacks in smart grid. IEEE Trans. Inf. Forensics Secur. **12**(7), 1735–1750 (2017)
49. J. Yao, P. Venkitasubramaniam, S. Kishore, L.V. Snyder, R.S. Blum, Network topology risk assessment of stealthy cyber attacks on advanced metering infrastructure networks, in *2017 51st Annual Conference on Information Sciences and Systems (CISS)*, pp. 1-6 (2017)
50. Z.-H. Yu, W.-L. Chin, Blind false data injection attack using PCA approximation method in smart grid. IEEE Trans. Smart Grid **6**(3), 1219–1226 (2015)

51. Y. Yuan, Z. Li, K. Ren, Quantitative analysis of load redistribution attacks in power systems. IEEE Trans. Parallel Distrib. Syst. **23**(9), 1731–1738 (2012)
52. Z. Zhang, R. Deng, D.K.Y. Yau, P. Cheng, J. Chen, On effectiveness of detecting FDI attacks on power grid using moving target defense, in *IEEE PES Innovative Smart Grid Technologies Conference (ISGT)*, pp. 1–5 (2019)
53. Z. Zhang, R. Deng, D.K.Y. Yau, P. Cheng, M.-Y. Chow, Security enhancement of power system state estimation with an effective and low-cost moving target defense. IEEE Trans. Syst. Man Cybern. Syst. **53**(5), 3066–3081 (2022)
54. R. Zhang, P. Venkitasubramaniam, Stealthy control signal attacks in linear quadratic gaussian control systems: detectability reward tradeoff. IEEE Trans. Inform. Forensics Secur. **12**(7), 1555–1570 (2017)
55. Z. Zhang, R. Deng, Y. Tian, P. Cheng, J. Ma, SPMA: stealthy physics-manipulated attack and countermeasures in cyber-physical smart grid. IEEE Trans. Inf. Forensics Secur. **18**, 581–596 (2022)
56. Z. Zhang, M. Sun, R. Deng, C. Kang, M.-Y. Chow, Physics-constrained robustness evaluation of intelligent security assessment for power systems. IEEE Trans. Power Syst. **38**(1), 872–884 (2023)
57. Z. Zhang, K. Yang, Y. Tian, J. Ma, An anti-disguise authentication system using the first impression of avatar in metaverse. IEEE Trans. Inf. Forensics Secur. **19**, 6393–6408 (2024)
58. R.D. Zimmerman, C.E. Murillo-Sanchez, R.J. Thomas, MATPOWER: steady-state operations, planning, and analysis tools for power systems research and education. IEEE Trans. Power Syst. **26**(1), 12–19 (2011)

Chapter 4
On Hiddenness of Moving Target Defense in the Smart Grid

4.1 Preliminary

In this section, we first introduce the DC power flow model adopted in this chapter. Then, we revisit the MTD used for defending FDI attacks in power grids.

4.1.1 System Model and False Data Injection Attack

System Model

We consider a power transmission network which consists of a set $\mathcal{N} = \{1, 2, \ldots, n+1\}$ of buses and a set $\mathcal{L} = \{k_1, k_2, \ldots, k_l\}$ of branches. Each bus $i \in \mathcal{N}$ is associated with a voltage phase angle θ_i. Each branch $k_t = \{i, j\} \in \mathcal{L}$ ($t \in \{1, 2, \ldots, l\}$) connects two buses i and j (from i to j). The susceptance of branch $\{i, j\}$ is b_{ij}. Suppose the power transmission network has a total of m monitored meters ($m > n$), and each branch is monitored by one meter. Let bus i' be the reference/slack bus. Then, the tth row of the measurement matrix $\mathbf{H} \in \mathbb{R}^{m \times n}$ is

$$\mathbf{H}_t = \begin{cases} [0 \ \ldots \ 0 \ \ -b_{ij} \ \ 0 \ \ldots \ 0 \ \ b_{ij} \ \ 0 \ \ldots 0], & \text{for } i \neq i', j \neq i' \\ [0 \ \ldots \ 0 \ \ \ 0 \ \ 0 \ \ldots \ 0 \ \ \pm b_{ij} \ \ 0 \ \ldots 0], & \text{for } i = i', j \neq i'. \end{cases} \quad (4.1)$$

We can see that $\mathbf{H}$ is a function of branch susceptances.

With the DC power flow model, the voltage phase angles are regarded as system states, which are denoted by $\mathbf{x} \in \mathbb{R}^n$. Assume that the measurements $\mathbf{z} \in \mathbb{R}^m$ are disturbed by i.i.d. noises $\mathbf{e} \in \mathbb{R}^m$, and e_i follows a Normal distribution with zero mean. Then the DC power flow model can be written as $\mathbf{z} = \mathbf{H}\mathbf{x} + \mathbf{e}$. To estimate the system states with the noisy measurements, there is an estimator $\hat{\mathbf{x}} = (\mathbf{H}^T\mathbf{W}\mathbf{H})^{-1}\mathbf{H}^T\mathbf{W}\mathbf{z} = \mathbf{K}\mathbf{z}$, where $\mathbf{K} = (\mathbf{H}^T\mathbf{W}\mathbf{H})^{-1}\mathbf{H}^T\mathbf{W}$ and $\mathbf{W}$ is a diagonal matrix whose elements are

R. Deng et al., *Moving Target Defense in the Smart Grid*, Wireless Networks,
https://doi.org/10.1007/978-3-031-92378-4_4

reciprocals of the noise variances. Since there may exist abnormal measurements, the BDD checker is widely used in SE. The BDD compares the Euclidean norm of the residues $\mathbf{r}$ between the measurements and their estimates with a predetermined threshold η, where $\mathbf{r} = \mathbf{z} - \hat{\mathbf{z}} = (\mathbf{I} - \mathbf{HK})\mathbf{z}$. The BDD alarm is triggered if $\|\mathbf{r}\|_2 > \eta$.

False Data Injection Attack

In this chapter, we focus on the FDI attacks against SE. We suppose that the malicious measurements are $\mathbf{z}_a = \mathbf{z} + \mathbf{a}$, where $\mathbf{a} \in \mathbb{R}^m$ denotes an attack vector. It has been proved that $\mathbf{z}_a$ cannot be detected by the BDD if $\mathbf{a} = \mathbf{Hc}$ [29], where $\mathbf{c} \in \mathbb{R}^n$ denotes the impacts on the state variables after FDI attacks. In other words, if the attack vector $\mathbf{a} \in S(\mathbf{H})$ holds, the malicious measurements $\mathbf{z}_a$ can undermine SE without being noticed, where $S(\cdot)$ denotes the (column) span of a matrix. For executing FDI attacks, we assume that the attacker has the following capabilities:

- The attacker is able to know partial or full information of the system topology and branch parameters through topology leaking attacks [37] or subspace attacks [53].
- The attacker is able to eavesdrop and tamper with the measurements by intruding into the communication network or IP-accessible field devices [47].

In real-world scenarios, there exist possibilities that attackers are able to execute eavesdropping attacks and tampering attacks on power grids. For example, NESCOR lists the failure scenarios that might happen in the AMI, DER, WAMPAC systems and so on [41]. It shows that the smart meter can be remotely accessed through installing malware (AMI.3), weak encryption used in the communication network might suffer from password cracking attack (AMI.24), measurements might be corrupted due to authentication compromise in data acquisition units (WAMPAC.4), and that the local measurement data sending by inter-substation link might be compromised (WAMPAC.11). Additionally, in Blackhat Europe 2014, Illera and Vidal [20] have shown that the smart meter can be compromised even it uses strong encryption. In GitHub, McIntyre [38] provided a platform for the security testing of smart meters, which may reveal potential vulnerabilities prone to be made use of by malicious users. Moreover, the BlackEnergy virus against Ukraine power grid also indicates that the attacker has the capability to eavesdrop and modify the measurements collected by the SCADA system [28]. From aforementioned possible examples of real-world attacks, the feasibility of our proposed attack model could be justified.

Moreover, we discuss the attack cost with respect to the efforts required to execute FDI attacks. The attacker needs to manipulate the raw measurements during their transmission to the control center. For example, he/she can corrupt the measurements in the field devices such as remote smart meters, RTUs and other IEDs, because these devices are usually lack of protection. Therefore, the attacker needs to know the locations and IP addresses of these field devices and the communication protocols used in them. Besides, in order to make the attack stealthy, the attacker should have enough information of the system topology and branch parameters. He/she needs to pay a lot of efforts and resources on learning this knowledge [37, 53]. Nevertheless, Liu et al. have investigated the design of FDI attacks with limited information of the system [31–33]. For example, the attacker only needs the topology and branch parameters within the attacking region for launching the load re-distribution attack [33].

Remark 4.1 In order to simplify the following discussions, we assume that the measurements are noiseless. That is, we have $\mathbf{z} = \mathbf{H}\mathbf{x}$. The corresponding estimates are given by $\hat{\mathbf{x}} = (\mathbf{H}^T\mathbf{H})^{-1}\mathbf{H}^T\mathbf{z} = \mathbf{x}$. The threshold η for the BDD is equal to 0. Nevertheless, we find that most results still hold when the measurements are noisy. In Sect. 4.6, we will evaluate the impacts of the measurement noise.

4.1.2 Moving Target Defense

Based on the deployment of D-FACTS (e.g., for the need of power flow control [12]), the system operator can actively perturb the branch parameters (i.e., admittance or impedance). Motivated by this use of D-FACTS, Morrow et al. [40] made the first attempt to perturb the branch impedance for thwarting FDI attacks. In recent studies [21, 24, 44], this defensive mechanism is termed as MTD. Here we briefly introduce how it works.

The basic idea of MTD is to avoid the attacker from anticipating the branch parameters by perturbing the branch impedance with D-FACTS. We assume that the control commands for activating D-FACTS can be protected from attackers. Let $\mathcal{L}_D$ be the set of branches that are deployed with D-FACTS devices. Once the MTD is activated, with the DC power flow model, the susceptances on branches in $\mathcal{L}_D$ are changed, that is,

$$b_{ij} \xrightarrow{\text{after MTD}} \widetilde{b}_{ij}, \text{ for } \{i, j\} \in \mathcal{L}_D, \tag{4.2}$$

where the arrow means that the branch susceptance is changed after MTD. According to Eq. (4.1), the measurement matrix $\mathbf{H}$ is changed accordingly. We denote the new measurement matrix as $\widetilde{\mathbf{H}}$. Assume that the attacker knows the measurement matrix $\mathbf{H}$. Then he/she can construct an attack vector in the form of $\mathbf{a} = \mathbf{H}\mathbf{c}$ with the attempt to bypass the BDD [29]. Correspondingly, the malicious measurements are $\mathbf{z}_a = \mathbf{z}_t + \mathbf{H}\mathbf{c}$, where $\mathbf{z}_t$ are the measurements after MTD. Thus, the Euclidean norm of the residues is

$$r_d(\mathbf{z}_a) = \|\mathbf{z}_t + \mathbf{H}\mathbf{c} - \widetilde{\mathbf{H}}\widetilde{\mathbf{K}}(\mathbf{z}_t + \mathbf{H}\mathbf{c})\| = \|\mathbf{a} - \widetilde{\mathbf{H}}\widetilde{\mathbf{K}}\mathbf{a}\|, \tag{4.3}$$

where $\widetilde{\mathbf{K}} = (\widetilde{\mathbf{H}}^T\widetilde{\mathbf{H}})^{-1}\widetilde{\mathbf{H}}^T$. Since $m > n$, we can prove that $\widetilde{\mathbf{H}}\widetilde{\mathbf{K}} \neq \mathbf{I}$, where $\mathbf{I}$ is an identity matrix of an appropriate size. As for $\mathbf{z}_a$, $r_d(\mathbf{z}_a) = 0$ if and only if $\mathbf{a} \in S(\widetilde{\mathbf{H}})$ [29]. However, since the attacker cannot anticipate $\widetilde{\mathbf{H}}$, it is possible that $\mathbf{a} \notin S(\widetilde{\mathbf{H}})$. That is, the FDI attacks can be detected by the BDD after MTD.

Therefore, the attacker fails to modify the state variables even the attack vector satisfies $\mathbf{a} = \mathbf{Hc}$. Based on this result, we present the condition needed for constructing an effective MTD.

Proposition 4.1 *An MTD is effective to thwart FDI attacks constructed with previous branch parameters only if there exists* $\boldsymbol{x} \in \mathbb{R}^n$ *such that* $\boldsymbol{x} \in S(\mathbf{H})$ *while* $\boldsymbol{x} \notin S(\widetilde{\mathbf{H}})$*, i.e.,* $S(\widetilde{\mathbf{H}}) \neq S(\mathbf{H})$.

Proof An MTD is effective means that the BDD after MTD can detect FDI attacks constructed with the previous branch parameters. In other words, there exists $\mathbf{a} = \mathbf{Hc}$ such that $r_d(\mathbf{z}_a) > 0$, where $\mathbf{z}_a = \mathbf{z}_t + \mathbf{a}$. Since $\mathbf{z}_t \in S(\widetilde{\mathbf{H}})$, we can derive that $\mathbf{a}$ must satisfy $\mathbf{a} \notin S(\widetilde{\mathbf{H}})$. It follows that there exists $\mathbf{x} \in \mathbb{R}^n$ such that $\mathbf{x} \in S(\mathbf{H})$ while $\mathbf{x} \notin S(\widetilde{\mathbf{H}})$. ■

Proposition 4.1 presents a necessary condition for achieving an effective MTD. It suggests that if $S(\widetilde{\mathbf{H}}) = S(\mathbf{H})$, then the MTD is not effective any more. We should emphasize that the effectiveness is the basic requirement of MTD regardless of the hiddenness.

4.2 Definition of Hiddenness

4.2.1 Hiddenness with Different Adversarial Knowledge

In this section, we first divide the attackers into two types according to their knowledge of the system. Then, we show that the hidden MTD proposed in [21] can be detected when the power flows are changed, which motivates us to investigate the sufficient and necessary condition for achieving a hidable MTD.

4.2.2 Detection of Moving Target Defense

We present two types of attackers according to their knowledge of the system. Their relationship can be specified in Fig. 4.1. Note that here we mainly discuss the attacker's capability to detect MTD. As for the design of FDI attacks after that, it is out of this chapter's scope.

Fully knowledgeable attackers. As the assumption given in [21], we assume that a fully knowledgeable attacker is able to know a measurement matrix $\mathbf{H}$ based on his/her understanding of the system, for example, through topology leaking attacks [37] or subspace attack [53]. Besides, the attacker is able to eavesdrop all measurements after MTD. Based on $\mathbf{H}$ and the measurements, the attacker can construct an MTD-confirming detector like the BDD for detecting MTD, which is given by

$$r_c(\mathbf{z}_t) = \|\mathbf{z}_t - \mathbf{H}(\mathbf{H}^T\mathbf{H})^{-1}\mathbf{H}^T\mathbf{z}_t\|. \tag{4.4}$$

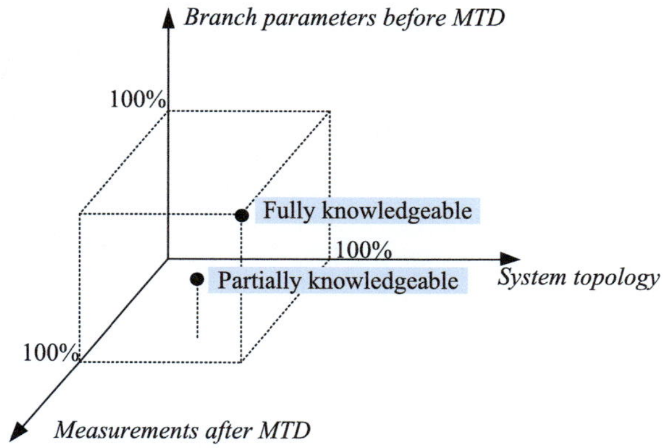

Fig. 4.1 The classification of attackers. Except the fully knowledgeable attacker, the others are partially knowledgeable attackers

where $\mathbf{z}_t$ are the measurements after MTD, i.e., $\mathbf{z}_t \in S(\widetilde{\mathbf{H}})$. Since $\mathbf{z}_t$ are generated corresponding to the new branch parameters, once $\mathbf{z}_t \notin S(\mathbf{H})$, the alarm of the MTD-confirming detector is triggered. Consequently, the MTD is detected by the attacker. However, since the power system is distributed in a large-scale geographically area, it is difficult for the attacker to obtain full information of the system topology, branch parameters and measurements. A more practical scenario is that the attacker is less knowledgeable.

Partially knowledgeable attackers. Compared to the fully knowledgeable attackers, a partially knowledgeable attacker has limited information of the system topology, the branch parameters before MTD and the measurements after MTD. Suppose the attacker can eavesdrop p measurements $\mathbf{z}_p$ ($\mathbf{z}_p \subseteq \mathbf{z}_t$) after MTD. Let $\mathbf{H}^p \in \mathbb{R}^{p\times n}$ be a submatrix of $\mathbf{H}$ constructed by the attacker corresponding to $\mathbf{z}_p$. Note that it is not necessary for the attacker to know the branch parameters just before MTD. The matrix $\mathbf{H}^p$ can be formed based on the attacker's understanding of the system. Then, we can derive a basis $\mathcal{B}$ of $S(\mathbf{H}^p)$ through row/column reduction of $\mathbf{H}^p$. Supposing the basis $\mathcal{B}$ forms a matrix $\mathbf{T}$, then we have $S(\mathbf{H}^p) = S(\mathbf{T})$. Thus, the attacker can construct an MTD-confirming detector like Eq. (4.4), which is given by

$$r_{pc}(\mathbf{z}_p) = \|\mathbf{z}_p - \mathbf{T}(\mathbf{T}^T\mathbf{T})^{-1}\mathbf{T}^T\mathbf{z}_p\|. \tag{4.5}$$

To construct such an MTD-confirming detector, the attacker only needs to know partial branch parameters before MTD and partial measurements after MTD. In that case, the complete information of all measurements and branch parameters is not necessary for detecting MTD. With this detector, we have $r_{pc}(\mathbf{z}_p) = 0$ if and only if $\mathbf{z}_p \in S(\mathbf{T})$ (i.e., $\mathbf{z}_p \in S(\mathbf{H}^p)$).

No matter whether the attacker is fully knowledgeable or partially knowledgeable, we assume that he/she launches an MTD confirming-first attack before FDI. Once the MTD is detected by the attacker, he/she will stop FDI and/or pay more efforts on the data exfiltration attacks for obtaining the latest branch parameters.

4.2.3 Comparison of Hidden Metrics

The detection of MTD is also exploited by Tian et al. [21], in which they proposed a hidden MTD to circumvent the fully knowledgeable attackers. But this hidden MTD is only hidden from the attacker when the power flows remain invariant. Here we give examples to illustrate its feasibility and limitation.

Given a 3-bus power system shown in Fig. 4.2, the branch susceptances are $b_{12} = -19.84$ p.u., $b_{13} = -17.48$ p.u. and $b_{23} = -15.72$ p.u.. Let the first bus be the reference bus. The corresponding measurement matrix is $\mathbf{H} = [-19.84, 0; 0, -17.48; 15.72, -15.72]$. The power injections of the buses are $P_{g_1} = 182$ MW, $P_{g_2} = 318$ MW, $P_{d_1} = 50$ MW, $P_{d_2} = 170$ MW and $P_{d_3} = 280$ MW. First, using the method proposed in [21], we perturb the branch susceptances as $\Delta b_{12} = -1.74$, $\Delta b_{13} = -6.64$ and $\Delta b_{23} = 1.50$ for guaranteeing that the power flows are not changed. From the third line of Table 4.1, we can see that $r_c(\mathbf{z}_t) = 0$ after MTD, which means that the hidden MTD cannot be detected by the MTD-confirming detector (4.4). Then, we change the system loads to $P_{d_1} = 50$ MW, $P_{d_2} = 150$ MW and $P_{d_3} = 300$ MW after the activation of this hidden MTD. The branch susceptances are the same as the perturbed results of the first step. From the fourth line of Table 4.1, we can see that the power flows are changed. Besides, as $r_c(\mathbf{z}_t) = 3.07 > 0$, the attacker can detect this hidden MTD. To prove this result further, we evaluate the hiddenness of the hidden MTD [21] in Sect. 4.6.3 using the real load data trace from New York State.

From above, we find that the proposed hidden MTD in [21] can be detected by an attacker when the power flows are changed. The detection of MTD may lead the attacker to give up FDI or pay more efforts and resources on learning the system's new

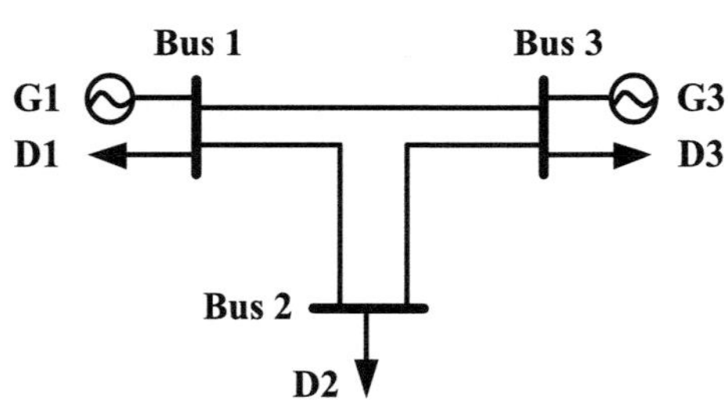

Fig. 4.2 The 3-bus power system

Table 4.1 Limitation of the hidden MTD proposed in [21]

Cases	Δb_{12}	Δb_{13}	Δb_{23}	z_{12}	z_{13}	z_{23}	$r_c(\mathbf{z}_t)$
Original	0	0	0	–107.26	–24.74	62.74	NaN
Hidden MTD	–1.74	–6.64	1.50	–107.26	–24.74	62.74	0
Load change	–1.74	–6.64	1.50	–101.61	–30.39	48.39	3.07

branch parameters. What's more, based on the detection result of MTD, the attacker might execute FDI attacks with their incomplete information of the system [7, 31]. Therefore, it is valuable to investigate the conditions needed for constructing a hidable MTD. Once the MTD cannot be detected by the attacker, the FDI attacks constructed with previous branch parameters can mostly be detected by the BDD after MTD, which increases the defender's chance to reverse and track the attacking routes. Before going deeper into the analysis, we present the definition of a hidable MTD.

Definition 4.1 A hidable MTD means that the MTD can never be detected by the attacker with his/her current knowledge.

For example, suppose that an attacker is fully knowledgeable. Then, the MTD is hidable only if all measurements $\mathbf{z}_t$ after MTD can bypass the MTD-confirming detector (4.4). That is, any $\mathbf{z}_t$ satisfies $\mathbf{z}_t \in S(\widetilde{\mathbf{H}})$ and $\mathbf{z}_t \in S(\mathbf{H})$. Once there exists $\mathbf{z}_t$ such that $\mathbf{z}_t \notin S(\mathbf{H})$, the MTD is not hidable. In the following, we will show that an MTD is possibly hidable only if the attacker is partially knowledgeable. With a fully knowledgeable attacker, we can never achieve a hidable and effective MTD by purely depending on the MTD design. To realize a hidable and effective MTD regardless of the changes of power flows and the attacker's knowledge, we need to jointly design the MPS and MTD.

4.3 System Model and Performance Analysis

4.3.1 Modeling Power Transmission Systems

Over a connected power transmission network, we define a set of branches as a *cyclic trunk* as follows.

Definition 4.2 Given a branch $k_t = \{i, j\}$, a cyclic trunk means that we can find a path along the other branches from j to i. The branch in the cyclic trunk is called cyclic branch.

For example, in the 3-bus power system shown in Fig. 4.2, there exists a cyclic trunk $1-3-2-1$. The branches $\{1, 2\}$, $\{1, 3\}$ and $\{2, 3\}$ are cyclic branches. In fact, not all branches in the power transmission network are cyclic branches. There exist branches that cannot be contained by any cyclic trunk. We define these branches as *open branches*.

Definition 4.3 Given a branch $k_t = \{i, j\}$, it is an open branch means that we can never find a path along the other branches from j to i.

For example, the 3-bus power system shown in Fig. 4.2 does not have open branch. While in the system given in Fig. 4.3, we find that there is only a single branch that connects *Subsystem_1* and *Subsystem_2*. Since this branch cannot be contained by any cyclic trunk, it is an open branch.

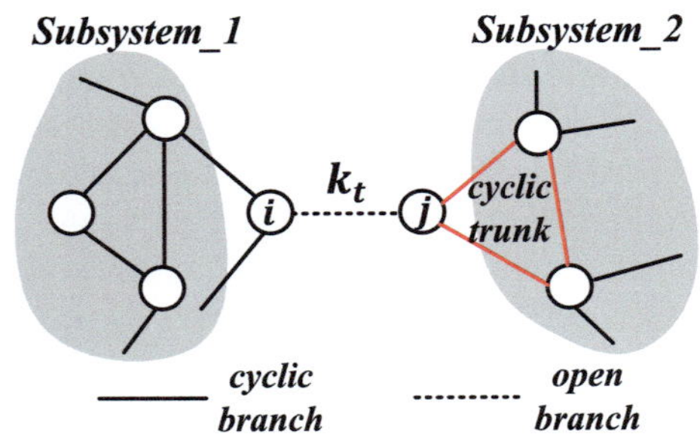

Fig. 4.3 The cyclic and open branches

4.3.2 Trade-Off Between Effectiveness and Hiddenness

Here we first present a sufficient and necessary condition for achieving a hidable MTD mathematically. Then, we explain this result with topology features such as the cyclic trunk, the cyclic branch and the open branch.

Mathematical Analysis

Let $\widetilde{\mathbf{H}}^p \in \mathbb{R}^{p\times n}$ be a submatrix of $\widetilde{\mathbf{H}}$ (after MTD) corresponding to $\mathbf{H}^p$ known by the attacker (see Sect. 4.2.2). The matrix $\mathbf{H}^p$ is constructed with previous branch parameters. Based on the MTD-confirming detector given by Eq. (4.5), we present a sufficient and necessary condition for achieving a hidable MTD.

Proposition 4.2 *An MTD is hidable if and only if* $S(\widetilde{\mathbf{H}}^p) \subseteq S(\mathbf{H}^p)$.

Proof (*Sufficiency*) Since $S(\widetilde{\mathbf{H}}^p) \subseteq S(\mathbf{H}^p)$, there exists $\mathbf{x} \in \mathbb{R}^n$ such that $\mathbf{H}^p\mathbf{x} = \widetilde{\mathbf{H}}^p\mathbf{x}_t$ for any $\mathbf{x}_t \in \mathbb{R}^n$. Suppose $\mathbf{z}_p = \widetilde{\mathbf{H}}^p\mathbf{x}_t$ are measurements after MTD. Then, we always have $\mathbf{x}$ such that $\mathbf{z}_p = \mathbf{H}^p\mathbf{x}$. It follows that $\mathbf{z}_p \in S(\mathbf{H}^p)$, which indicates that $\mathbf{z}_p$ can always bypass the MTD-confirming detector (i.e., $r_{pc}(\mathbf{z}_p) = 0$). Therefore, the MTD is hidable.

(Necessity) We resort to a contradiction. That is, suppose an MTD is hidable while $S(\widetilde{\mathbf{H}}^p) \nsubseteq S(\mathbf{H}^p)$. The later means that there exists $\mathbf{x}_t \in \mathbb{R}^n$ such that $\widetilde{\mathbf{H}}^p\mathbf{x}_t \neq \mathbf{H}^p\mathbf{x}$ for any $\mathbf{x} \in \mathbb{R}^n$. Suppose $\mathbf{z}_p = \widetilde{\mathbf{H}}^p\mathbf{x}_t$ are measurements after MTD. Then, we have $\mathbf{z}_p \notin S(\mathbf{H}^p)$. It follows that $r_{pc}(\mathbf{z}_p) \neq 0$, which means that the MTD can be detected by the MTD-confirming detector. This contradicts to the original assumption. In that case, we must have $S(\widetilde{\mathbf{H}}^p) \subseteq S(\mathbf{H}^p)$ for achieving a hidable MTD. ■

Proposition 4.2 implies that an MTD is hidable if and only if the measurements corresponding to $\widetilde{\mathbf{H}}^p$ after MTD can always bypass the MTD-confirming detector (4.5). If $p = m$, we have $S(\widetilde{\mathbf{H}}) \subseteq S(\mathbf{H})$. For deeper analysis, we give an equivalent criteria to $S(\widetilde{\mathbf{H}}^p) \subseteq S(\mathbf{H}^p)$.

Lemma 4.1 *Let* $S(\mathbf{A})$ *and* $S(\widetilde{\mathbf{A}})$ *be the (column) span of matrices* $\mathbf{A} \in \mathbb{R}^{m\times n}$ *and* $\widetilde{\mathbf{A}} \in \mathbb{R}^{m\times n}$, *respectively.* $S(\mathbf{A}) \subseteq S(\widetilde{\mathbf{A}})$ *if and only if there exists a matrix* $\mathbf{C} \in \mathbb{R}^{n\times n}$ *such that* $\mathbf{A} = \widetilde{\mathbf{A}}\mathbf{C}$.

Proof (*Sufficiency*) The (column) span of $\widetilde{\mathbf{A}}$ is $S(\widetilde{\mathbf{A}}) = \{\mathbf{u} \in \mathbb{R}^m | \mathbf{u} = \widetilde{\mathbf{A}}\mathbf{v} = \sum_{j=1}^{n} v_j \tilde{\mathbf{a}}_j,\ v_j \in \mathbf{v}, \mathbf{v} \in \mathbb{R}^n\}$, where $\tilde{\mathbf{a}}_j$ is the jth column of $\widetilde{\mathbf{A}}$. The span of $\widetilde{\mathbf{A}}\mathbf{C}$ is $S(\widetilde{\mathbf{A}}\mathbf{C}) = \{\mathbf{u}' \in \mathbb{R}^m | \mathbf{u}' = \widetilde{\mathbf{A}}\mathbf{C}\mathbf{v}' = \sum_{j=1}^{n} v_j'(\sum_{i=1}^{n} c_{ij}\tilde{\mathbf{a}}_i),\ v_j' \in \mathbf{v}, \mathbf{v} \in \mathbb{R}^n\}$, where c_{ij} is the element at the position (i, j) of matrix $\mathbf{C}$. Since $\sum_{i=1}^{n} c_{ij}\tilde{\mathbf{a}}_i \in S(\widetilde{\mathbf{A}})$ for any $1 \le j \le n$, we have $\mathbf{u}' = \sum_{j=1}^{n} v_j'(\sum_{i=1}^{n} c_{ij}\tilde{\mathbf{a}}_i) \in S(\widetilde{\mathbf{A}})$ according to the property of the subspace (the column span of a matrix is a subspace) [19]. Thus, we can derive that $S(\widetilde{\mathbf{A}}\mathbf{C}) \subseteq S(\mathbf{A})$. Since $\mathbf{A} = \widetilde{\mathbf{A}}\mathbf{C}$, we have $S(\mathbf{A}) \subseteq S(\widetilde{\mathbf{A}})$.

(Necessity) Since $S(\mathbf{A}) \subseteq S(\widetilde{\mathbf{A}})$, for any $\mathbf{v}'' \in \mathbb{R}^n$, there exists $\mathbf{v} \in \mathbb{R}^n$ such that $\mathbf{A}\mathbf{v}'' = \widetilde{\mathbf{A}}\mathbf{v}$. Suppose $\mathbf{v}_i''$ is a vector with "1" in the ith position and zeros elsewhere. Then, the matrix $[\mathbf{v}_1''\ \mathbf{v}_2''\ \cdots\ \mathbf{v}_n'']$ is an n by n identity matrix. There must exist $\mathbf{v}_1, \mathbf{v}_2, \ldots, \mathbf{v}_n$ such that $\mathbf{A}\mathbf{v}_1'' = \widetilde{\mathbf{A}}\mathbf{v}_1, \mathbf{A}\mathbf{v}_2'' = \widetilde{\mathbf{A}}\mathbf{v}_2, \ldots, \mathbf{A}\mathbf{v}_n'' = \widetilde{\mathbf{A}}\mathbf{v}_n$. It follows that $\mathbf{A}[\mathbf{v}_1''\ \mathbf{v}_2''\ \cdots\ \mathbf{v}_n''] = \widetilde{\mathbf{A}}[\mathbf{v}_1\ \mathbf{v}_2\ \cdots\ \mathbf{v}_n]$. Thus, we have $\mathbf{A} = \widetilde{\mathbf{A}}[\mathbf{v}_1\ \mathbf{v}_2\ \cdots\ \mathbf{v}_n] = \widetilde{\mathbf{A}}\mathbf{C}$, where $\mathbf{C}$ is an n by n matrix formed by $\mathbf{v}_1, \mathbf{v}_2, \ldots, \mathbf{v}_n$. ■

Lemma 4.1 implies that if there exists a matrix $\mathbf{Y} \in \mathbb{R}^{n \times n}$ such that $\widetilde{\mathbf{H}}^p = \mathbf{H}^p\mathbf{Y}$, then the MTD is hidable. In the following, we associate this mathematical result with the topological features of the power transmission network.

4.4 Analysis of Hiddenness of Moving Target Defense

4.4.1 General Analysis and Physical Understanding

Based on the mathematical result, we analyze the hiddenness and effectiveness of MTD when we only perturb the cyclic and open branches, respectively. One thing we should emphasize here is that an MTD must be effective to thwart FDI attacks first rather than being hidable.

Perturbing cyclic branches. According to the definition, every cyclic branch is contained by at least one cyclic trunk. If the perturbed branches are cyclic branches, we can derive the following result.

Theorem 4.1 *An MTD is hidable only if the branch susceptances in the cyclic trunk, which contains the perturbed branches, are perturbed with a same ratio.*

Proof We renumber the buses and branches for simplification. The cyclic trunk which contains the perturbed branches is denoted by $1 \to 2 \to \cdots \to p \to 1$, where p is the number of buses in this cyclic trunk. The set of branches contained by the cyclic trunk is denoted by $\mathcal{L}^c = \{k_1, k_2, \ldots, k_p\}$. The corresponding susceptances are $\{b_{k_1}, b_{k_2}, \ldots, b_{k_p}\}$. Corresponding to this cyclic trunk, we construct a submatrix $\mathbf{H}^c$ of $\mathbf{H}$ as

$$\mathbf{H}^c = \begin{bmatrix} -b_{k_1} & b_{k_1} & 0 & 0 & \cdots & 0 & 0 \cdots 0 \\ 0 & -b_{k_2} & b_{k_2} & 0 & \cdots & 0 & 0 \cdots 0 \\ \vdots & \vdots & \vdots & \vdots & \vdots & \vdots & \vdots\ \vdots\ \vdots \\ 0 & \cdots & 0 & \cdots & -b_{k_{p-1}} & b_{k_{p-1}} & 0 \cdots 0 \\ b_{k_p} & \cdots & 0 & \cdots & 0 & -b_{k_p} & 0 \cdots 0 \end{bmatrix}. \tag{4.6}$$

The matrix $\mathbf{H}^c$ is a p by n matrix. Its rank is $p-1$. We can see that the first $p-1$ columns form a basis of the column space of $\mathbf{H}^c$. This basis forms a matrix

$$\mathbf{T}^c = \begin{bmatrix} -b_{k_1} & b_{k_1} & 0 & 0 & \cdots & 0 \\ 0 & -b_{k_2} & b_{k_2} & 0 & \cdots & 0 \\ \vdots & \vdots & \vdots & \vdots & \vdots & \vdots \\ 0 & \cdots & 0 & \cdots & 0 & -b_{k_{p-1}} \\ b_{k_p} & \cdots & 0 & \cdots & 0 & 0 \end{bmatrix}. \tag{4.7}$$

The matrix $\mathbf{T}^c$ is a p by $p-1$ full column rank matrix. After MTD, suppose that the susceptances of branches in $\mathcal{L}^c$ are $\{\widetilde{b}_{k_1}, \widetilde{b}_{k_2}, \ldots, \widetilde{b}_{k_p}\}$, where $\widetilde{b}_{k_1} = \lambda_1 b_{k_1}$, $\widetilde{b}_{k_2} = \lambda_2 b_{k_2}$, ..., $\widetilde{b}_{k_p} = \lambda_p b_{k_p}$, where $\lambda_i > 0$ is the perturbation ratio of branch k_i. Corresponding to $\mathbf{H}^c$, the submatrix is $\widetilde{\mathbf{H}}^c$ of $\widetilde{\mathbf{H}}$ after MTD. Like $\mathbf{H}^c$, the first $p-1$ columns form a basis of $\widetilde{\mathbf{H}}^c$. Besides, the matrix formed by the basis is

$$\widetilde{\mathbf{T}}^c = \begin{bmatrix} -\widetilde{b}_{k_1} & \widetilde{b}_{k_1} & 0 & 0 & \cdots & 0 \\ 0 & -\widetilde{b}_{k_2} & \widetilde{b}_{k_2} & 0 & \cdots & 0 \\ \vdots & \vdots & \vdots & \vdots & \vdots & \vdots \\ 0 & \cdots & 0 & \cdots & 0 & -\widetilde{b}_{k_{p-1}} \\ \widetilde{b}_{k_p} & \cdots & 0 & \cdots & 0 & 0 \end{bmatrix}. \tag{4.8}$$

We rewrite $\mathbf{T}^c$ and $\widetilde{\mathbf{T}}^c$ as $\mathbf{T}^c = \begin{bmatrix} \mathbf{U}^c \\ \mathbf{L}^c \end{bmatrix}$ and $\widetilde{\mathbf{T}}^c = \begin{bmatrix} \widetilde{\mathbf{U}}^c \\ \widetilde{\mathbf{L}}^c \end{bmatrix}$, respectively. The matrices $\mathbf{U}^c$ and $\widetilde{\mathbf{U}}^c$ are both $p-1$ by $p-1$ invertible matrices. Besides, we have $\mathbf{L}^c = \begin{bmatrix} b_{k_p} & \cdots & 0 & \cdots & 0 & 0 \end{bmatrix}$ and $\widetilde{\mathbf{L}}^c = \begin{bmatrix} \widetilde{b}_{k_p} & \cdots & 0 & \cdots & 0 & 0 \end{bmatrix}$.

Since the MTD is hidable, according to the Proposition 4.2, we have $S(\widetilde{\mathbf{T}}^c) \subseteq S(\mathbf{T}^c)$. Further, with Lemma 4.1, there exists a matrix $\mathbf{G} \in \mathbb{R}^{(p-1)\times(p-1)}$ such that $\widetilde{\mathbf{T}}^c = \mathbf{T}^c\mathbf{G}$. It follows that $\widetilde{\mathbf{U}}^c = \mathbf{U}^c\mathbf{G}$. Using $\widetilde{\mathbf{U}}^c$ and $\mathbf{U}^c$, we can derive that

$$\mathbf{G} = \begin{bmatrix} \lambda_1 & \lambda_2 - \lambda_1 & \cdots & \lambda_{p-1} - \lambda_{p-2} \\ 0 & \lambda_2 & \cdots & \lambda_{p-1} - \lambda_{p-2} \\ \vdots & \vdots & \vdots & \vdots \\ 0 & \cdots & \lambda_{p-2} & \lambda_{p-1} - \lambda_{p-2} \\ \cdots & 0 & 0 & \lambda_{p-1} \end{bmatrix}.$$

Moreover, we have $\widetilde{\mathbf{L}}^c = \mathbf{L}^c\mathbf{G}$. Thus, we can derive that $\lambda_2 - \lambda_1 = 0$, $\lambda_3 - \lambda_2 = 0$, ..., $\lambda_{p-1} - \lambda_{p-2} = 0$. Therefore, the matrix $\mathbf{G}$ becomes

$$\mathbf{G} = \begin{bmatrix} \lambda_1 & 0 & 0 & 0 & \cdots & 0 \\ 0 & \lambda_1 & 0 & 0 & \cdots & 0 \\ \vdots & \vdots & \vdots & \vdots & \vdots & \vdots \\ 0 & \cdots & 0 & \cdots & \lambda_1 & 0 \\ 0 & \cdots & 0 & \cdots & 0 & \lambda_1 \end{bmatrix} = \lambda_1 \mathbf{I}_{(p-1)\times(p-1)} \tag{4.9}$$

where $\mathbf{I}_{(p-1)\times(p-1)}$ is a $p-1$ by $p-1$ identity matrix. Therefore, we have $\lambda_1 = \lambda_2 = \cdots = \lambda_p$. ∎

Theorem 4.1 suggests that, to make an MTD hidable, we must perturb all branch susceptances in a cyclic trunk with a same ratio. In other words, we cannot achieve a hidable MTD by randomly perturbing the cyclic branches.

Remark 4.2 To the contrary, from the attacker's perspective, as long as the attacker is able to know a cyclic trunk that contains the perturbed branches and construct a corresponding MTD-confirming detector as (4.5), the MTD can be detected if the branch susceptances in this cyclic trunk are not perturbed with a same ratio with respect to the values that the attacker has known.

Therefore, to thwart the attacker's capability to detect MTD, the defender needs to perturb as many cyclic branches as possible with a same ratio. For example, given a 4-bus power system as Fig. 4.4, the perturbed branch k_1 is contained by three cyclic trunks, i.e., $\{k_1, k_2, k_5\}$, $\{k_1, k_4, k_5\}$ and $\{k_1, k_2, k_3, k_4\}$. To achieve a hidable MTD, all branch susceptances in this system should be perturbed with a same ratio. That is, only perturbing k_1 cannot guarantee that the MTD is hidable. However, in that case, we find that the MTD is not effective anymore. This issue is discussed in the following.

Given a power transmission network as Fig. 4.3, *Subsystem_1* and *Subsystem_2* are only connected by an open branch $k_t = \{i, j\}$. Suppose that the branches $\mathcal{L}_1 = \{k_1, k_2, \ldots, k_{t-1}\}$ in *Subsystem_1* and the branches $\mathcal{L}_2 = \{k_{t+1}, k_{t+2}, \ldots, k_l\}$ in *Subsystem_2* are all cyclic branches. We analyze the hiddenness and effectiveness of MTD when all branch susceptances in both subsystems are perturbed with a same ratio.

We rearrange $\mathbf{H}$ as $\mathbf{H} = \left[\mathbf{H}^u \ \mathbf{h}_i \ \mathbf{h}_j \ \mathbf{H}^v\right]$, where $\mathbf{H}^u \in \mathbb{R}^{m\times n_1}$ and $\mathbf{H}^v \in \mathbb{R}^{m\times n_2}$ correspond to *Subsystem_1* and *Subsystem_2*, respectively, n_1 and n_2 are the numbers of buses in these two subsystems and $n_1 + n_2 + 2 = n$. Similarly, we rearrange $\widetilde{\mathbf{H}}$ (after

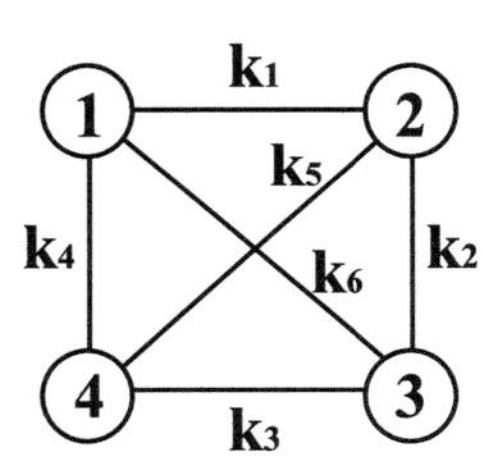

Fig. 4.4 A 4-bus power system with six branches

MTD) as $\widetilde{\mathbf{H}} = \left[\widetilde{\mathbf{H}}^u \ \widetilde{\mathbf{h}}_i \ \widetilde{\mathbf{h}}_j \ \widetilde{\mathbf{H}}^v\right]$. Since the branch susceptances in both subsystems are perturbed with a same ratio with respect to the values in $\mathbf{H}$, we have $\widetilde{\mathbf{H}}^u = \lambda_u \mathbf{H}^u$ and $\widetilde{\mathbf{H}}^v = \lambda_v \mathbf{H}^v$, where $\lambda_u > 0$ and $\lambda_v > 0$. Suppose the reference/slack bus is in the *Subsystem_1*. Let $\mathbf{h}_i$ and $\widetilde{\mathbf{h}}_i$ be the ith column of $\mathbf{H}$ and $\widetilde{\mathbf{H}}$, respectively. According to Eq. (4.1), if we add all columns in the combined matrix $\left[\mathbf{h}_j \ \mathbf{H}^v\right]$ together, we can obtain a vector as

$$\mathbf{h} = [\ 0 \ \ldots \ 0 \quad \underbrace{b_{ij}}_{k_t{}^{\text{th}} \text{ element}} \quad 0 \ \ldots \ 0]^T, \tag{4.10}$$

where the k_tth element is the only non-zero element in $\mathbf{h}$. Since bus i is connected to buses in *Subsystem_1*, the elements in $\widetilde{\mathbf{h}}_i$ are λ_u times of the corresponding elements in $\mathbf{h}_i$ except the k_tth element $-b_{ij}$. It follows that $\widetilde{\mathbf{h}}_i = \lambda_u \mathbf{h}_i + (-1 + \lambda_u)\mathbf{h}$. Similarly, we have $\widetilde{\mathbf{h}}_j = \lambda_v \mathbf{h}_j + (1 - \lambda_v)\mathbf{h}$. Therefore, there exists a matrix $\mathbf{X} \in \mathbb{R}^{n \times n}$ such that $\widetilde{\mathbf{H}} = \mathbf{H}\mathbf{X}$, where $\mathbf{X}$ is given by

$$\mathbf{X} = \left[\mathbf{X}_1 \ \mathbf{X}_2 \ \mathbf{X}_3\right], \mathbf{X}_1 = \begin{bmatrix} \lambda_u \mathbf{I}_{n_1 \times n_1} \\ \mathbf{0}_{(n-n_1) \times n_1} \end{bmatrix}, \mathbf{X}_2 = \begin{bmatrix} \mathbf{0}_{n_1 \times 1} & \mathbf{0}_{n_1 \times 1} \\ \lambda_u & 0 \\ -1 + \lambda_u & \lambda_v \\ -1 + \lambda_u & 1 - \lambda_v \\ \vdots & \vdots \end{bmatrix}, \mathbf{X}_3 = \begin{bmatrix} \mathbf{0}_{(n-n_2) \times n_2} \\ \lambda_v \mathbf{I}_{n_2 \times n_2} \end{bmatrix}, \tag{4.11}$$

where $\mathbf{I}_{n_1 \times n_1}$ is an n_1 by n_1 identity matrix, $\mathbf{0}_{(n-n_1) \times n_1}$ is an $n - n_1$ by n_1 zero matrix. Since the determinant of $\mathbf{X}$ is $\lambda_u^{n_1+1} \lambda_v^{n_2+1} \neq 0$, the matrix $\mathbf{X}$ is invertible. Thus, we can derive that $S(\widetilde{\mathbf{H}}) = S(\mathbf{H})$ according to the Lemma 4.1. It means that, even though the MTD can be hidable when we perturb all branch susceptances in the subsystems with a same ratio, the MTD is not effective anymore (see Proposition 4.1). That is, the BDD after MTD cannot detect any FDI attack constructed with the previous measurement matrix $\mathbf{H}$. It is trivial to realize a hidable MTD when the MTD is not useful.

Perturbing open branches. According to the definition, an open branch is not contained by any cyclic trunk. As for perturbing open branches only, we obtain the following result.

Theorem 4.2 *An MTD is hidable but not effective to thwart FDI attacks constructed with previous branch parameters if only the open branches are perturbed.*

Proof Suppose that only the open branch $k_t = \{i, j\}$ is perturbed to $\widetilde{b}_{ij} = \lambda b_{ij}$ in the power transmission network shown in Fig. 4.3, where $\lambda > 0$. According to Eq. (4.10), we can derive that $\widetilde{\mathbf{h}}_i = \mathbf{h}_i + (-\lambda + 1)\mathbf{h}$ and $\widetilde{\mathbf{h}}_j = \mathbf{h}_j + (\lambda - 1)\mathbf{h}$. Therefore, there must exist a matrix $\mathbf{X}$ such that $\widetilde{\mathbf{H}} = \mathbf{H}\mathbf{X}$, where

$$\mathbf{X} = \begin{bmatrix} \mathbf{X}_1 \ \mathbf{X}_2 \ \mathbf{X}_3 \end{bmatrix}, \mathbf{X}_1 = \begin{bmatrix} \mathbf{I}_{n_1 \times n_1} \\ \mathbf{0}_{(n-n_1)\times n_1} \end{bmatrix}, \mathbf{X}_2 = \begin{bmatrix} \mathbf{0}_{n_1\times 1} & \mathbf{0}_{n_1 \times 1} \\ 1 & 0 \\ -\lambda + 1 & 1 \\ -\lambda + 1 & \lambda - 1 \\ \vdots & \vdots \end{bmatrix}, \mathbf{X}_3 = \begin{bmatrix} \mathbf{0}_{(n-n_2)\times n_2} \\ \mathbf{I}_{n_2 \times n_2} \end{bmatrix}. \tag{4.12}$$

Because the determinant of $\mathbf{X}$ is 1, the matrix $\mathbf{X}$ is invertible. Therefore, we have $S(\widetilde{\mathbf{H}}) = S(\mathbf{H})$, which indicates that the MTD is not effective to thwart FDI attacks. ■

Theorem 4.2 implies that, it is trivial to perturb the open branches only for constructing MTD.

Discussion on Effectiveness of Moving Target Defense

According to the forgoing analysis, we find that an MTD is no more effective under two cases: (i) we perturb all branch susceptances in a system or a subsystem that is separated by open branches with a same ratio; (ii) we only perturb the open branches. Therefore, to guarantee the effectiveness of MTD, we should not perturb the branch susceptances like the abovementioned two cases.

4.4.2 Hiddenness with Partially Knowledgeable Attackers

Next, we exploit the system information needed for a partially knowledgeable attacker to detect MTD. Based on Theorem 4.1, we first analyze whether an MTD is hidable or not when the attacker is unable to know a cyclic trunk.

Theorem 4.3 *An MTD is hidable if the branches whose susceptances are known by the attacker cannot form a cyclic trunk that contains the perturbed branches.*

Proof Assume that the attacker knows a set of branches $\mathcal{L}^a = \{k_1, k_2, \ldots, k_p\}$. Based on $\mathcal{L}^a$, the attacker constructs a submatrix $\mathbf{H}^a$ of $\mathbf{H}$. First, consider that $\mathcal{L}^a$ cannot form a cyclic trunk. This case can be regarded as that the power transmission network only contains "open branches" in the eyes of the attacker. Thus, it is the same problem as that we have analyzed in Theorem 4.2. Let $\widetilde{\mathbf{H}}^a$ be a submatrix of $\widetilde{\mathbf{H}}$ (after MTD). The matrix $\widetilde{\mathbf{H}}^a$ is formed corresponding to $\mathbf{H}^a$. Then, we always have $S(\mathbf{H}^a) = S(\widetilde{\mathbf{H}}^a)$ regardless of the perturbed branches. Therefore, the MTD is hidable. Second, consider that $\mathcal{L}^a$ forms a cyclic trunk but it does not contain the perturbed branches. We have $S(\mathbf{H}^a) = S(\widetilde{\mathbf{H}}^a)$ since $\mathbf{H}^a = \widetilde{\mathbf{H}}^a$. Therefore, the MTD is hidable. ■

Theorem 4.3 implies that if the branches known by the attacker cannot form a cyclic trunk containing the perturbed branches, any MTD cannot be detected by the attacker.

Remark 4.3 Based on the above result, we find an additional case where the MTD is hidable as well. That is, the attacker is able to know a cyclic trunk that contains the perturbed branches but cannot construct an MTD-confirming detector corresponding to it. For example, the attacker is unable to eavesdrop some measurements corresponding to the branches in the cyclic trunk that he/she knows (see Sect. 4.2.2). This result inspires us to jointly design the MPS and MTD for achieving a hidable MTD in Sect. 4.5.

Furthermore, we derive that a partially knowledgeable attacker must have the following system information to detect MTD.

Theorem 4.4 *An MTD can be detected by a partially knowledgeable attacker only if: (i) the set of branches known by him/her can form a cyclic trunk that contains the perturbed branches; (ii) the measurements corresponding to the branches in this cyclic trunk are known by him/her for constructing an MTD-confirming detector.*

Proof We resort to a contradiction. That is, suppose that an MTD can be detected by the attacker while at least one of the abovementioned conditions is not satisfied. For the first condition, we have proved that an MTD cannot be detected if the attacker is unable to know a cyclic trunk that contains the perturbed branches in Theorem 4.3. This contradicts to the original assumption. For the second condition, if some measurements on branches in the known cyclic trunk cannot be obtained by the attacker, then he/she is unable to construct a corresponding MTD-confirming detector to detect the MTD, which contradicts to the original assumption as well. Therefore, if at least one of the conditions is not satisfied, the attacker cannot detect the MTD, i.e., the MTD is hidable. ■

Theorem 4.4 presents the system information that the attacker must have for detecting MTD. Since it is impossible for the defender to know the attacker's capability, there may always exist perturbed branches that are contained in the cyclic trunks which are known by the attacker, resulting in that the MTD is detected. Nevertheless, Theorem 4.4 motivates us to realize a hidable MTD by combining MPS with MTD. We present this MPS+MTD co-design strategy in Sect. 4.5.

4.4.3 Hiddenness with Fully Knowledgeable Attackers

Since the power transmission network contains cyclic branches and/or open branches, we can classify it into three types: (i) containing cyclic branches only; (ii) containing cyclic branches and open branches; (iii) containing open branches only. In the following, we discuss the hiddenness of MTD under these three types of power transmission networks, respectively. Unfortunately, in order to guarantee the effectiveness of MTD, we observe that any MTD cannot be hidable under the fully knowledgeable attackers.

Type (i). Since the power transmission network is formed by cyclic branches only, the only way to make an MTD hidable under the fully knowledgeable attackers is to perturb all branch susceptances with a same ratio. However, we have proved that the corresponding MTD is not effective anymore.

Type (ii). For the type (ii) power transmission network, there are two ways to make an MTD hidable. The first way is to perturb all branch susceptances in the subsystem that is separated by open branches with a same ratio. The second way is to perturb the open branches only. However, we have proved that, even though the MTD is hidable with these two ways, the MTD is not effective anymore.

Type (iii). Since we have proved that the MTD is not effective anymore when we only perturb the open branches, it is impossible to make an MTD effective and hidable under the fully knowledgeable attackers in this type of power transmission network.

4.5 Improved Hidden Moving Target Defense

According to the forgoing analysis, we cannot make sure that an MTD is effective and hidable by purely depending on the MTD design, because it is impossible to obtain the attacker's knowledge. Once the MTD is detected, a conservative attacker may give up launching FDI attacks. While an aggressive attacker may spend more resources on learning the latest branch parameters and/or launch FDI attacks with his/her limited information of the system. In the later scenario, the system might still face the risk of being attacked. Therefore, it is necessary and important to design a hidable and effective MTD. Since a hidable MTD is stealthy to the attacker, the FDI attacks can mostly be detected by the BDD after MTD. After that, the attacks can be redirected to a honeypot for further analysis, which increases the defender's chance to expose the details of the attacking routes.

Basically, a hidable MTD should be hidden from the attacker regardless of the changes of power flows, the attacker's knowledge and the branch perturbations. In this chapter, we jointly design MTD and MPS to achieve this goal. We term this MPS+MTD co-design strategy as the *improved hidden MTD*. The main idea of the improved hidden MTD is to prevent the attacker from knowing some measurements after the activation of MTD. Thus, we can thwart the attacker's capability to construct the MTD-confirming detector.

According to Theorem 4.4, an MTD can be detected by the attacker only if he/she is able to construct an MTD-confirming detector corresponding to a cyclic trunk that contains the perturbed branches. According to Sect. 4.2.2, the construction of an MTD-confirming detector not only needs the branch parameters but also the measurements on the corresponding branches. Once the measurements are protected, the attacker cannot construct an MTD-confirming detector based on the branch parameters only. In that case, we can make an MTD hidable.

Corollary 4.1 *If we protect the measurements on some branches such that the attacker cannot construct an MTD-confirming detector corresponding to any cyclic trunk that contains the perturbed branches, then the MTD is hidable.*

Corollary 4.1 provides a way to combine MPS with MTD for realizing a hidable MTD. It is said that we can make the MTD stealthy to the attacker by thwarting the attacker's capability to obtain the measurements corresponding to the cyclic trunks that contain the perturbed branches. Therefore, it seems that we should find all cyclic trunks that have perturbed branches, which is easy with small-sized power systems, because we can traverse all branches and determine whether the set of selected branches can form a circle or not. But it is computational intensive when the size of the power transmission network is large. On the other hand, one may find that it is not necessary to find all the related cyclic trunks. An easy way to realize a hidable MTD is just to protect the measurements on the perturbed branches. However, this method has some shortcomings. First, since different perturbed branches may be contained by the same set of cyclic trunks, some protected measurements are not necessary. For example, in the 4-bus power system shown in Fig. 4.5, the perturbed branches are $\{1, 2\}$ and $\{2, 3\}$. Thus, the protected measurement on the branch $\{1, 2\}$ (or $\{2, 3\}$) is not necessary for a hidable MTD. Second, since the protected measurements are exactly the same as the perturbed branches, a careful attacker might be able to infer the deployment of MTD through these protections. Third, it is not necessary to deploy these two protection mechanisms in the same place. They should complement each other and create more limitations for the attacker. For example, to protect the state variable of bus 2 in the 4-bus power system (Fig. 4.5), we can perturb the parameters of branch $\{1, 2\}$ and $\{2, 3\}$ or protect the measurements on branch $\{1, 2\}$ and $\{2, 3\}$. MPS should increase the difficulty for the attacker to construct FDI attacks after MTD. For instance, if we perturb the parameters of branch $\{1, 2\}$ and $\{2, 3\}$ and protect the measurements on branch $\{3, 4\}$, then not only the MTD is hidable, but also the attacker cannot construct feasible FDI attacks to modify the state variables of all buses. To address these issues, we propose an MPS+MTD co-design strategy to achieve a hidable MTD based on the graph theory.

The MPS+MTD co-design strategy is based on the insight that: a spanning tree of a connected graph does not contain any circle, while an unique circle is formed when we add an additional edge to the spanning tree. We explain this insight in our context. Suppose that the set of perturbed branches is $\mathcal{L}_D$. These branches cover a set of interconnected buses $\mathcal{N}_D$. Then, based on the power transmission network, we can construct a subnetwork $\mathcal{G}_D$ by separating the set of buses $\mathcal{N}_D$ from the outside. Assume that the set of branches $\mathcal{T}_D$ forms a spanning tree of $\mathcal{G}_D$. Then, in $\mathcal{G}_D$, if we protect the measurements on the rest branches (except $\mathcal{T}_D$), the attacker cannot construct an MTD-confirming detector corresponding to any cyclic trunk

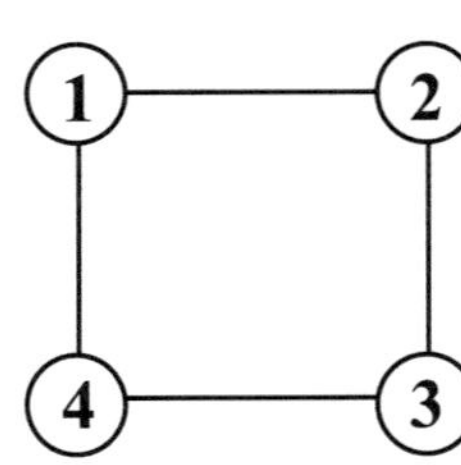

Fig. 4.5 A 4-bus power system with four branches

Algorithm 3: Find a group of interconnected buses and its spanning tree

Data: $\mathcal{N}_D, \mathcal{G}$

Initialization: $s_0 = s \in \mathcal{N}_D, \mathcal{U} \leftarrow \emptyset, \mathcal{T} \leftarrow \emptyset, \mathcal{V} \leftarrow \emptyset$, $stop = 0$;
Construct an undirected graph $\mathcal{G}_D$ according to $\mathcal{N}_D$ and the power network $\mathcal{G}$;
$\forall i \in \mathcal{N}_D$, compute the set of buses C_i connected to bus i according to $\mathcal{G}_D$;
while $stop \neq 1$ **do**
    if *All buses in* C_s *belong to* $\mathcal{V}$ **then**
        $s = \mathcal{U}(s)$;
    else
        for *each* $j \in C_s$ **do**
            if $j \notin \mathcal{V}$ **then**
                put (s, j) into $\mathcal{T}$, put j into $\mathcal{V}$, $\mathcal{U}(j) = s$, $s = j$;
                break;
            end
        end
    end
    if $s = s_0$ **then**
        $stop = 1$;
    end
end

Result: $\mathcal{V}, \mathcal{T}$

within $\mathcal{G}_D$. Next, one more concern is that partial of the spanning tree $\mathcal{T}_D$ that contains the perturbed branches might form cyclic trunks with the outside branches. To address this concern, we protect the measurements on branches that connect $\mathcal{G}_D$ to the outside. To the end, the attacker can never construct an MTD-confirming detector corresponding to any cyclic trunk that contains the perturbed branches. That is, the MTD is hidable. The main steps for designing the improved hidden MTD are as follows:

1. Given a set of perturbed branches $\mathcal{L}_D$, the set of covered buses form a set $\mathcal{N}_D$. We use Algorithm 3 to find a group of interconnected buses $\mathcal{V}$ and its spanning tree $\mathcal{T}$. The computational complexity of Algorithm 3 is $O(n_d + l_d + 1)$, where n_d is the number of buses and l_d is the number of branches in $\mathcal{G}_D$. Using this algorithm recursively, if there exists, we can obtain several disjoint groups of buses $\mathcal{V}_D^1$, $\mathcal{V}_D^2, \ldots$ and their spanning trees $\mathcal{T}_D^1, \mathcal{T}_D^2, \ldots$.
2. Based on $\mathcal{V}_D^i$ and $\mathcal{T}_D^i$, we find the set of protected measurements $\mathcal{P}_D^i$ in $\mathcal{G}_D^i$, where $\mathcal{G}_D^i$ is a subnetwork formed according to $\mathcal{V}_D^i$ and the power transmission network $\mathcal{G}$. The set $\mathcal{P}_D^i$ corresponds to the set of branches in $\mathcal{G}_D^i$ that are not in $\mathcal{T}_D^i$.
3. Moreover, for each $\mathcal{G}_D^i$, we find the set of border buses $\mathcal{V}_{BD}^i$, which are connected to the outside. For each border bus, we divide its branches that connect to the outside buses into the cyclic and open branches. If all these branches are open branches, we remove this border bus from the border bus set. After that, the border bus set $\mathcal{V}_{BD}^i$ is updated.

4. We select $|\mathcal{V}^i_{BD}| - 1$ border buses and protect the measurements corresponding to the cyclic branches that connect to the outside buses. The set of protected measurements is $\mathcal{P}^i_{BD}$.
5. Finally, for each $\mathcal{G}^i_D$, we obtain the set of protected measurements $\mathcal{P}^i_D + \mathcal{P}^i_{BD}$.

From above, step 3 and step 4 are based on the fact that the open branches are not contained by any cyclic trunk. The measurements on the open branches are not necessarily protected for a hidable MTD. Moreover, the reason why we only choose $|\mathcal{V}^i_{BD}| - 1$ border buses rather than all border buses is because, the cyclic trunks, which are formed by partial of the spanning tree $\mathcal{T}^i_D$ and several outside branches, must get in $\mathcal{N}^i_D$ from one border bus and get out from another border bus. By protecting the measurements on the cyclic branches, which connected the border buses to outside, of $|\mathcal{V}^i_{BD}| - 1$ border buses, it is enough to thwart the attacker's capability to detect the MTD.

In summary, after we have determined the set of perturbed branches for constructing MTD, the set of measurements that must be protected for a hidable MTD is computed according to steps (1)–(5). The selection of perturbed branches for constructing an effective MTD has been discussed in our previous work [50]. Here we mainly focus on the realization of a hidable MTD. For the MPS+MTD co-design strategy, we present a simple example here. As shown in Fig. 4.6, the MTD is constructed by perturbing branches $\{4, 5\}$, $\{4, 9\}$, $\{5, 6\}$ and $\{7, 8\}$. They cover the buses 4, 5, 6, 7, 8 and 9. We can see that the covered buses are interconnected with each other. The corresponding graph is formed by nodes 4, 5, 6, 7, 8 and 9 and edges $\{4, 5\}$, $\{4, 9\}$, $\{5, 6\}$, $\{6, 7\}$, $\{7, 8\}$ and $\{8, 9\}$. Its spanning tree can be constructed as $6 - 5 - 4 - 9 - 8 - 7$ [step (1)]. Thus, in order to achieve a hidable MTD, the measurement on branch $\{6, 7\}$ should be protected [step (2)]. The border buses of this subnetwork/graph are 4, 6, 7 and 8. Since the branches $\{1, 4\}$ and $\{2, 8\}$ are open branches, we only need to protect the measurements on the cyclic branches that connect bus 6 and 7 to the outside [step (3)]. It follows that we should protect the measurement on branch $\{7, 10\}$ or $\{3, 6\}$. The figure shows that we protect the

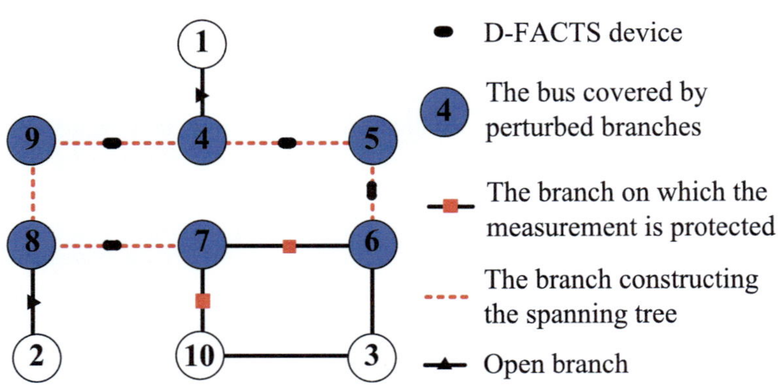

Fig. 4.6 An example of 10-bus power system

measurement on branch {3, 6} [step (4)]. Therefore, the measurements of branches {6, 7} and {3, 6} should be protected for realizing a hidable MTD [step (5)]. More examples are presented in Sect. 4.6.4.

Theorem 4.5 *Given a power transmission network $\mathcal{G}$, if we protect the measurements on branches except those in the spanning tree $\mathcal{T}$ of $\mathcal{G}$, then any MTD is hidable.*

Proof According to the property of the spanning tree of a graph, any additional edge/branch added to the spanning tree $\mathcal{T}$ can form a circle. Therefore, if we protect the measurements on branches except those in the spanning tree $\mathcal{T}$ of the power transmission network $\mathcal{G}$, the attacker can never construct an MTD-confirming detector corresponding to any cyclic trunk in $\mathcal{G}$. Thus, any MTD is hidable. ■

Theorem 4.5 suggests that we could jointly design MPS and MTD to achieve a hidable MTD regardless of the perturbed branches and the attacker's knowledge. For example, in the IEEE 9-bus power system shown in Fig. 4.7, we can construct a spanning tree of the power transmission network using branches {1, 4}, {4, 9}, {9, 8}, {8, 2}, {8, 7}, {7, 6}, {6, 3} and {6, 5}. The rest branch is {4, 5}. In that case, if we protect the measurement on branch {4, 5}, the attacker can never construct an MTD-confirming detector corresponding to any cyclic trunk regardless of the perturbed branches, which means that any MTD is hidable.

Discussion. In fact, MPS not only can help making the MTD hidable, but also can add more constraints for the attacker to construct FDI attacks after MTD. For example, suppose the set of protected measurements is $\mathcal{M}$. The rows of the measurement matrix $\widetilde{\mathbf{H}}$ corresponding to $\mathcal{M}$ form a submatrix $\widetilde{\mathbf{H}}^M$. In that case, if the attacker wants to design an attack vector as $\mathbf{a} = \mathbf{Hc}$ to bypass the BDD after MTD, the following equation must be satisfied

$$\mathbf{a}^M = \widetilde{\mathbf{H}}^M\mathbf{c} = \mathbf{0}, \tag{4.13}$$

where the vector $\mathbf{a}^M$ corresponds to the protected measurements, $\mathbf{0}$ is a zero vector of an appropriate size. Therefore, apart from the constraints introduced by MTD,

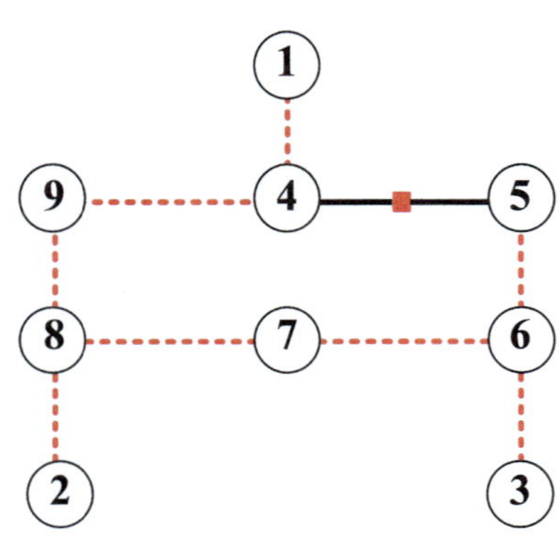

Fig. 4.7 IEEE 9-bus power system

the attacker must consider the constraints given in Eq. (4.13). This indicates that the benefits of the MPS+MTD co-design strategy are twofold: making the MTD hidable and enhancing the security of SE further. Besides, since MPS is mainly used for realizing hidable MTD, the number of protected measurements sometimes is small (for example, only one measurement is protected in the 9-bus power system mentioned above).

4.6 Performance Evaluation

In this section, the hiddenness and effectiveness of MTD are evaluated with the IEEE 14-bus (Fig. 4.8), 57-bus and 118-bus test power systems. For the IEEE 14-bus power system, we inject a load data trace from New York State in one day (09-07-2018) every five minutes [42] for simulations. All simulations are carried out in MATLAB. We assume that all measurements have the same noise parameter, which is a normal distribution $\mathcal{N}(0, \sigma^2)$. The BDD threshold η is set as $\sigma\sqrt{\chi^2_{m-n,\alpha}}$, where $m-n$ is the freedom degree of the Chi-square distribution and α is the false alarm rate, which is 0.05. In practice, the branch susceptance must be within some limits, i.e., $b_{ij}^{min} \leq b_{ij} \leq b_{ij}^{max}$. Since b_{ij} is a negative value, we set $b_{ij}^{min} = (1+pm)b_{ij}$ and $b_{ij}^{max} = b_{ij}/(1+pm)$, where pm is the parameter of the susceptance perturbation range [21, 45]. pm is usually set as 0.2, but 0.5 is also tolerable. Here we vary pm from 0.05 to 0.5 spaced with 0.05.

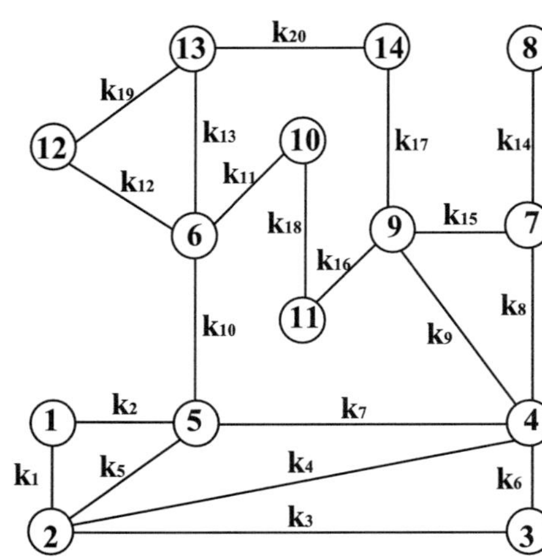

Fig. 4.8 The IEEE 14-bus power system

4.6.1 *Effectiveness and Hiddenness of Moving Target Defense*

Here we consider the effectiveness and hiddenness of MTD with respect to the branch perturbation range. We construct the FDI attack vector $\mathbf{a} = \mathbf{Hc}$ by sampling the value of c_i from a uniform distribution $\mathcal{U}(-1, 1)$. The noise variance is fixed as 0.01. The attacker is assumed to be fully knowledgeable. We adopt the IEEE 14-bus power system. For each setting (e.g., $pm = 0.2$), we conduct simulations with the load data trace in the day 09-07-2018 of New York State.

First, we verify that the MTD is not effective to thwart FDI attacks anymore if we perturb the cyclic branches in the subsystem (separated by the open branch k_{14}) with a same ratio. The simulation result is shown in Fig. 4.9. We can see that the detection probability of MTD is lower than 2%, which can be ignored compared to the false alarm rate 5%. Therefore, the MTD is hidable. While the detection probability of the FDI attacks is around 5%, which approaches the false alarm rate. It indicates that the BDD after MTD misses detecting almost all FDI attacks. The simulation result highlights that even though the MTD is hidable, it is not effective to thwart FDI attacks anymore, regardless of the perturbation range and the load change. Second, we perturb the open branch k_{14} only. The simulation result is also plotted in Fig. 4.9. We can see that the result is the same with the first case. The detection probability of MTD and the detection probability of FDI attacks are both negligible, i.e., around 5%. The simulation result indicates that, even though the MTD is hidable by perturbing the open branch only, it is not effective to thwart FDI attacks anymore, regardless of the perturbation range and the load change. By contrast, we randomly perturb the susceptances of cyclic branches k_1, k_2, k_3 and k_4. The detection probabilities of MTD and FDI attacks are plotted in Fig. 4.9. We can see that both of them increase with the perturbation range. The simulation results show that we cannot achieve a hidable and effective MTD simultaneously by purely depending on the MTD-design.

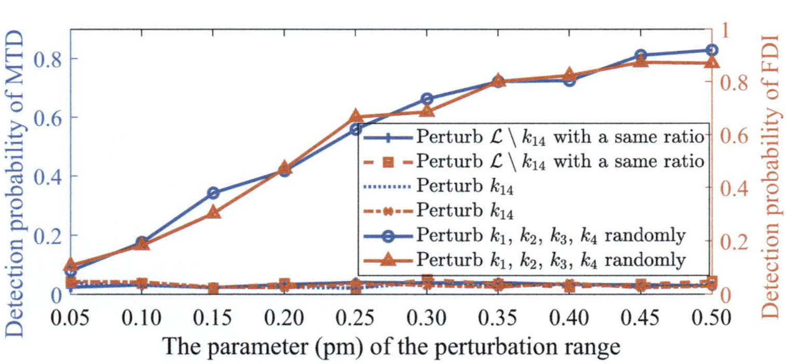

Fig. 4.9 The detection probability of MTD and FDI attacks when purely depending on the MTD design

4.6.2 Hiddenness with Partially Knowledgeable Attackers

Next, we evaluate the hiddenness of MTD when the attacker is partially knowledgeable. According to Sect. 4.2.2, the submatrix $\mathbf{H}^p$ may not be full row/column rank. Thus, for a noisy system, we determine the threshold η for an MTD-confirming detector by calculating $\sigma\sqrt{\chi^2_{p-r(\mathbf{H}^p),\alpha}}$, where p is the number of rows of $\mathbf{H}^p$ and $r(\mathbf{H}^p)$ is the rank of $\mathbf{H}^p$. In simulations of this section, the noise variance is fixed as 0.01. We adopt the IEEE 14-bus power system. For each setting, we conduct simulations with the load data trace from New York State on 09-07-2018.

To begin with, we evaluate the hiddenness of MTD with several cases in which the attacker has different knowledge. In all cases, only the branch k_7 is perturbed. In the first case, we assume that the attacker has known the previous susceptances of the set $\mathcal{L}^1 = \{k_5, k_6, k_7, k_8\}$ of branches. Besides, he/she is able to eavesdrop the measurements z_5, z_6, z_7 and z_8. We can see that the perturbed branch k_7 is contained in $\mathcal{L}^1$, but $\mathcal{L}^1$ cannot form a cyclic trunk. From Fig. 4.10, we can see that the detection probability of MTD approaches 0 in this case, which means that the MTD is always hidable regardless of the perturbation range and the load change. In the second case, we assume that the attacker has known the previous susceptances of the set $\mathcal{L}^2 = \{k_3, k_5, k_6, k_7\}$ of branches, but he/she cannot eavesdrop the measurement z_5. Therefore, even though $\mathcal{L}^2$ can form a cyclic trunk, the attacker cannot construct an MTD-confirming detector corresponding to it. From Fig. 4.10, we can see that the detection probability of MTD approaches 0 as well, which means that the MTD is hidable regardless of the perturbation range and the load change. In the third case, we assume that the attacker has known the previous susceptances of the set $\mathcal{L}^3 = \{k_1, k_2, k_3, k_4, k_5, k_6\}$ of branches, in which $\{k_1, k_2, k_5\}$ and $\{k_3, k_4, k_6\}$ can form a cyclic trunk, respectively. However, the branch k_7 is not contained in both cyclic trunks. From Fig. 4.10, we can see that the detection probability of MTD approaches 0 in this case, which means that the MTD is hidable regardless of the perturbation range and the load change. The above cases show that the attacker is unable to detect

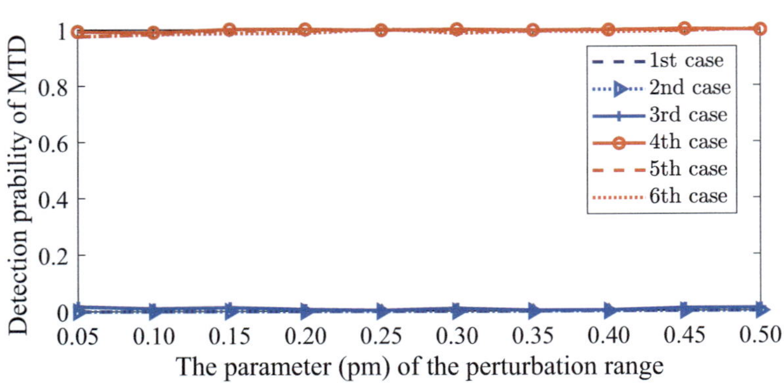

Fig. 4.10 The detection probability of MTD with respect to the attacker's knowledge

the MTD with his/her current knowledge. By contrast, the following cases have the opposite results. We assume that attacker has known the previous susceptances of the set $\mathcal{L}^4 = \{k_4, k_5, k_7\}$ in the fourth case, the set $\mathcal{L}^5 = \{k_3, k_5, k_6, k_7\}$ of branches in the fifth case and $\mathcal{L}^6 = \{k_7, k_9, k_{10}, k_{11}, k_{16}, k_{18}\}$ of branches in the sixth case. Besides, the attacker is able to eavesdrop the measurements on branches corresponding to $\mathcal{L}^4$, $\mathcal{L}^5$ and $\mathcal{L}^6$, respectively. We can see that branch k_7 is contained by $\mathcal{L}^4$, $\mathcal{L}^5$ and $\mathcal{L}^6$. $\mathcal{L}^4$, $\mathcal{L}^5$ and $\mathcal{L}^6$ all can form a cyclic trunk. Moreover, the attacker is able to construct the MTD-confirming detectors corresponding to them. From Fig. 4.10, we can see that the detection probabilities of MTD all approach 100% in the later three cases. The simulation results are consistent with our analysis in Sect. 4.4.

Further, the hiddenness of MTD is evaluated under the partially knowledgeable attacker when the branches are perturbed in different ways. We assume that the attacker has known the previous susceptances of the set $\mathcal{L}^a = \{k_4, k_5, k_7\}$ of branches and is able to eavesdrop the corresponding measurements z_4, z_5 and z_7. We can see that $\mathcal{L}^a$ forms a cyclic trunk. Besides, the attacker is able to construct an MTD-confirming detector corresponding to it. As a comparison, we perturb the branches in $\mathcal{L}^a$ with a same ratio and randomly, respectively. The simulation results are plotted in Fig. 4.11. We can see that the detection probability of MTD approaches 0 when the branches in $\mathcal{L}^a$ are perturbed with a same ratio, which means that the MTD is hidable regardless of the perturbation range and the load change. While the detection probability of MTD is almost 100% when we perturb the branches in $\mathcal{L}^a$ randomly, which means that the MTD can be detected by the attacker regardless of the load change. The simulation results suggest that we might realize a hidable MTD by perturbing the susceptances of some (not all) branches with a same ratio, but it depends on the knowledge of the attacker.

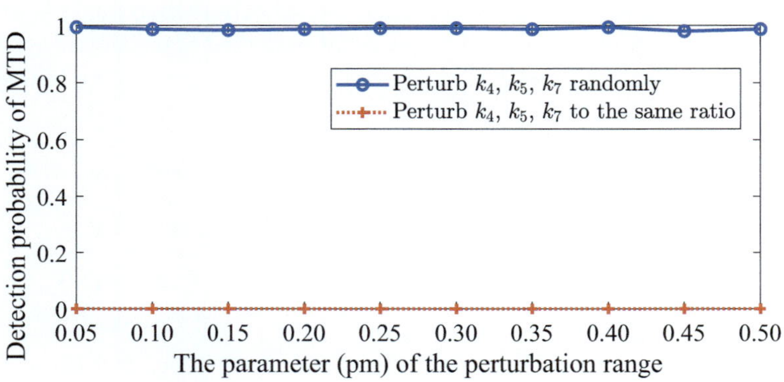

Fig. 4.11 Detection probability of MTD vs the branch perturbations of a cyclic trunk

4.6.3 Hiddenness with Fully Knowledgeable Attackers

With the basic requirement that an MTD must be effective, we verify that any MTD is not hidable under the fully knowledgeable attackers.

Impact of Perturbation Range and Measurement Noise

First, the hiddenness of MTD is evaluated with respect to the perturbation range. In the IEEE 14-bus power system, we perturb the branch k_1 and k_2. They are perturbed within the same perturbation range at each time. The parameter of the perturbation range is tested from 0.05 to 0.5. For each setting, we conduct simulations with the load data trace from New York State on 09-07-2018. The simulation results are plotted in Fig. 4.12. For the noiseless measurements, we can see that the detection probability of MTD is 100%, which means that the MTD can always be detected by the attacker regardless of the perturbation range and the load change. For the noisy measurements, we can see that even though the detection probability of MTD is affected by the parameter of the noise, it increases with the perturbation range. For example, the detection probability of MTD is 95% when $pm = 0.20$ and $\sigma^2 = 0.01$, while it is 100% when $pm = 0.50$ and $\sigma^2 = 0.01$.

Impact of Number of Perturbed Branches

Next, we validate that we can never achieve a hidable MTD under the fully knowledge attacker even we change the number of perturbed branches. The IEEE 14-bus power system is adopted. The parameter of the perturbation range is fixed as $pm = 0.2$. For each setting (e.g., the number of perturbed branches is 4), we conduct simulations with the load data trace from New York State on 09-07-2018. The

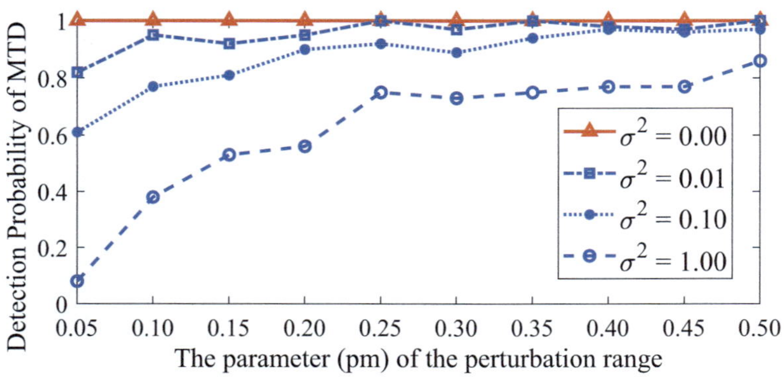

Fig. 4.12 Detection probability of MTD vs the perturbation range and measurement noise

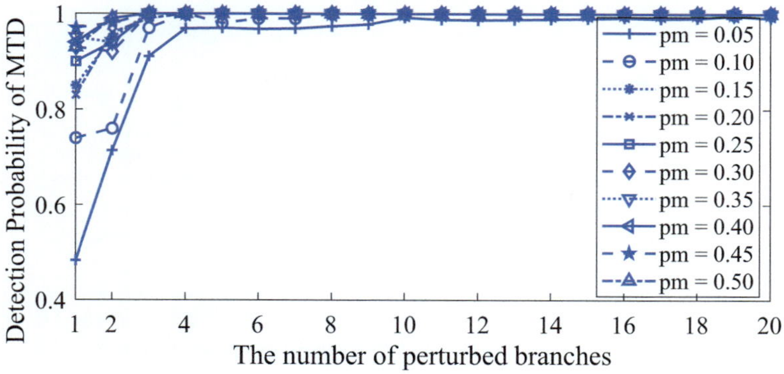

Fig. 4.13 Detection probability of MTD vs different numbers of perturbed branches

noise variance is set to be 0.01. We perturb the set of branches successively as $\{k_1\}$, $\{k_1, k_2\}$, $\{k_1, k_2, k_3\}$, ... The simulation results are plotted in Fig. 4.13. We can see that, with more branches being perturbed, the detection probability of MTD is larger. The detection probability of MTD is over 90% when we perturb the set of branches $\{k_1, k_2, k_3, k_4, k_5, k_6\}$. The simulation results demonstrate that we cannot achieve a hidable MTD under the fully knowledgeable attacker with different numbers of branches being perturbed.

Impact of System Size

Moreover, under the fully knowledgeable attacker, we validate that we cannot realize a hidable MTD with the IEEE 57-bus and 118-bus power system as well. We perturb the susceptances of the first and second branch specified in MATPOWER [54]. Based on the default settings and data, we inject random errors within [–10, 10] into their loads. The simulation results are plotted in Fig. 4.14. When the measurements are noiseless, we can see that the detection probability of MTD is 100%, which means that the MTD can always be detected by the attacker regardless of the size of the power system and the load change. As for noisy measurements (the noise variance is 0.01), we can see that the detection probability of MTD is affected by the noise. But for both power systems, the detection probabilities of MTD increase with the perturbation range.

4.6.4 Improved Hidden Moving Target Defense

Here we evaluate the hiddenness and effectiveness of the improved hidden MTD proposed in Sect. 4.5. With the IEEE 14-bus power system, we conduct simulations with the load data trace from New York State on 09-07-2018.

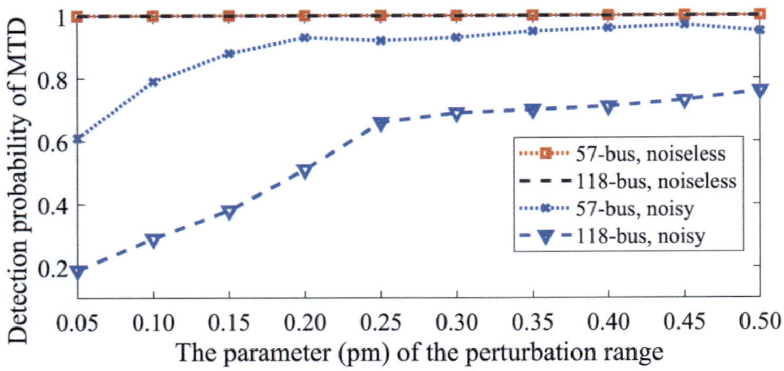

Fig. 4.14 Detection probability of MTD with the IEEE 57-bus and 118-bus power systems

Co-design of Measurement Protection Strategy and Moving Target Defense

We realize the improved hidden MTD by jointly design MPS and MTD with the IEEE 14-bus power system. We perturb the branches $k_{10}, k_{12}, k_{13}, k_{18}$ and k_{19}. They cover a set {6, 9, 10, 11, 12, 13, 14} of buses. Using Algorithm 3, we can see that these buses are interconnected with each other. Thus, Algorithm 3 outputs a spanning tree of the corresponding subnetwork, which is formed by branches $k_{11}, k_{12}, k_{13}, k_{16}, k_{17}$ and k_{20}. Based on the steps given in Sect. 4.5, we find that the protected measurements should be z_{10}, z_{18} and z_{19}. The result is plotted in Fig. 4.15. We can see that the number of protected measurements (i.e., 3) is less than the number of perturbed branches (i.e., 5), and the protected measurements are not on the perturbed branches.

Moreover, we consider the co-design of MTD and MPS with the large-scale power system. We take the IEEE 118-bus power system as an example. To construct MTD,

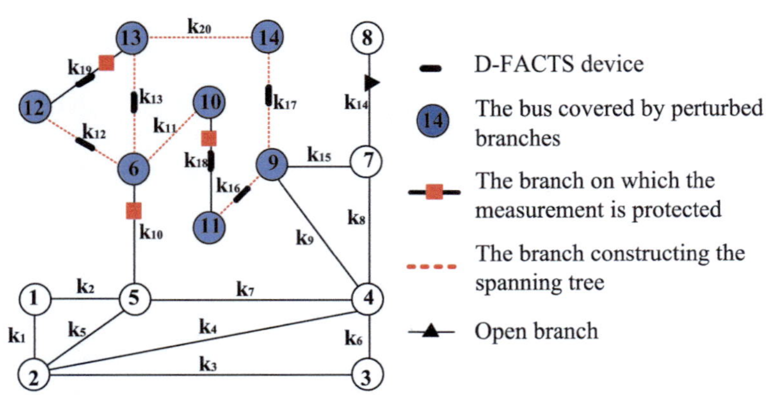

Fig. 4.15 The deployment of the MPS+MTD co-design strategy

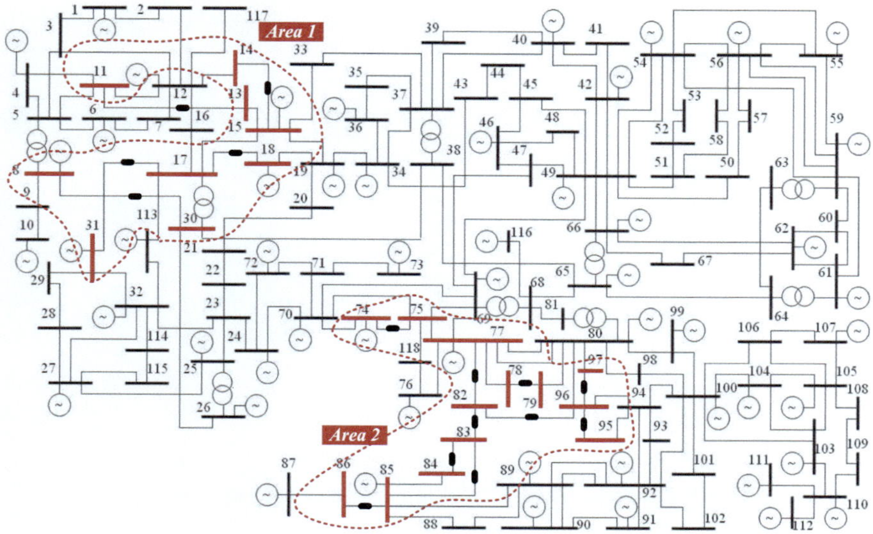

Fig. 4.16 The construction of MTD with the IEEE 118-bus power system

Fig. 4.17 Co-design of MTD and MPS: Area 1

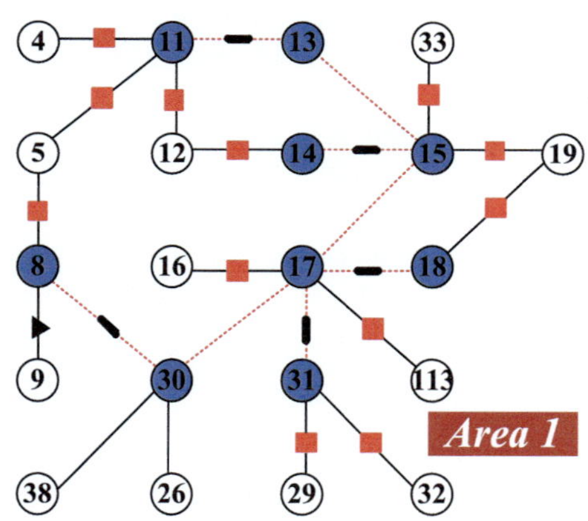

the perturbed branches are shown in Fig. 4.16. With Algorithm 1, we find that the covered buses by these perturbed branches can be divided into two areas, i.e., area 1 and area 2. In each area, the buses are interconnected with each other. We plot the transmission network of area 1 and area 2 in Figs. 4.17 and 4.18, respectively. The spanning trees of the corresponding graphs/subnetworks of area 1 and area 2 are denoted with red dashed line. According to steps (1)–(5) in Sect. 4.5, we mark the branches on which the measurements should be protected in Figs. 4.17 and 4.18. Note that the measurements of the open branches are not necessarily protected. Besides, the measurements on the cyclic branches that connect one of the border buses to the outside are not necessarily protected as well. The result indicates that the improved hidden MTD can be scaled to large-scale power systems.

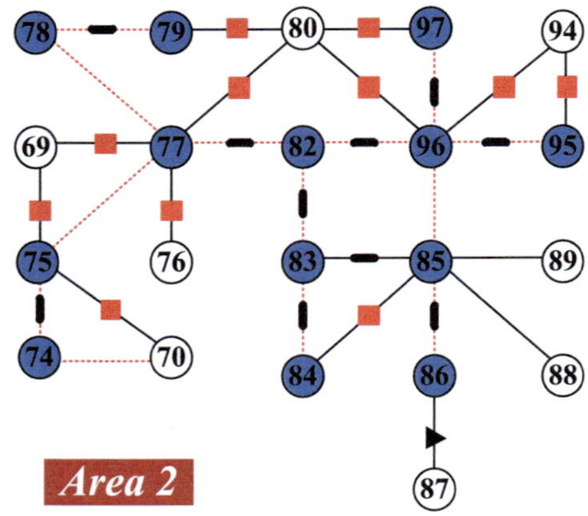

Fig. 4.18 Co-design of MTD and MPS: Area 2

Comparison with Existing Hidden Moving Target Defense

Next, we compare the effectiveness and hiddenness of the improved hidden MTD with the hidden MTD [21] and the random MTD [44]. Based on the algorithm given in [21], we need to change the state variables of bus 13 and 14 and perturb the branch k_{13}, k_{17}, k_{19} and k_{20} to achieve the hidden MTD. The susceptance perturbation ratios (i.e., $\widetilde{b_{ij}}/b_{ij}$) of k_{13}, k_{17}, k_{19} and k_{20} are 0.97, 0.99, 0.82 and 1.02. While to realize an improved hidden MTD, we need to protect the measurements z_{10}, z_{11} and z_{19}. The perturbation ranges for the improved hidden MTD and the random MTD are both set to be 0.2. The deployments of MTD and MPS are shown in Fig. 4.19.

Effectiveness. The effectiveness of MTD is measured by the probability of unsuccessful FDI attacks in an attacking pool after MTD. The attacking pool is generated as follows. First, the state bias c_i is identically sampled from the uniform distri-

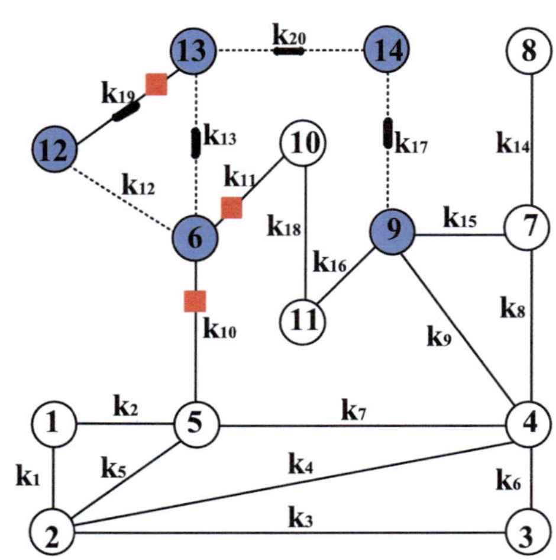

Fig. 4.19 The deployment of the hidden MTD [21] (the state variables of bus 13 and 14 are changed) and the MPS+MTD strategy

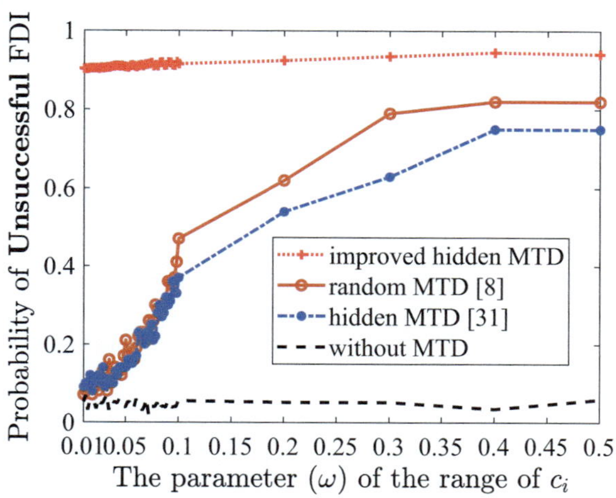

Fig. 4.20 The effectiveness to thwart FDI attacks with improved hidden MTD, random MTD [44], hidden MTD [21] and the BDD without MTD

bution $\mathcal{U}(-\omega, \omega)$ ($\omega > 0$). Then, for the attack vector $\mathbf{a} = \mathbf{Hc}$, the modified state variables (i.e., the non-zero elements in $\mathbf{c}$) are uniformly selected. For each setting (i.e., $\omega = 0.1$), we generate 1000 attack vectors, i.e., the size of the attacking pool is 1000. The probability of unsuccessful FDI attacks (Pr) is defined as $\text{Pr} = \frac{\text{\# of unsuccessful attack vectors}}{1000}$. The unsuccessful attack vectors include those which cannot bypass the BDD after MTD and those which manipulate the protected measurements. Here we adopt the IEEE 14-bus power system. The default settings and data from MATPOWER are used. The standard deviation of the measurement noise is 0.01. We plot the simulation results in Fig. 4.20. We can see that the probability of unsuccessful FDI attacks with the improved hidden MTD is above 90% regardless of the range of c_i, especially when the range of c_i is small (less than [–0.1,0.1]). While the probability of unsuccessful FDI attacks with the hidden MTD [21] and the random MTD [44] are smaller but increase with the range of c_i. In the later two methods, when the range of c_i is $[-0.1, 0.1]$, the probability of unsuccessful FDI attacks after MTD is less than 50%. That is, when the range of c_i is smaller, the probability of unsuccessful FDI attacks is smaller. The above simulation result indicates that the improved hidden MTD performs better than the hidden MTD [21] and random MTD [44].

Hiddenness. We assume that the attacker is fully knowledgeable. To avoid confusion, corresponding to the improved hidden MTD, the fully knowledgeable attacker knows the measurements that are not protected after MTD. In the IEEE 14-bus power system, we keep track of the hiddenness of MTD using the load data trace from New York State on 09-07-2018. To see the variations of power flows in this day, we plot the relative changes (compared to the start point) of the power flows in Fig. 4.21. The simulation results about the hiddenness of MTD are plotted in Fig. 4.22. From Fig. 4.21, we can divide the changes of power flows into two periods. During the first period, the power flows change within 20% mostly. While in the second period, the power flows change dramatically, which is over 50% sometimes. From Fig. 4.22, we

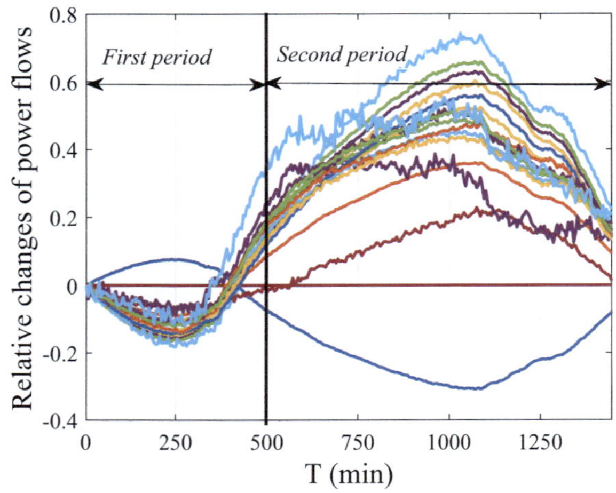

Fig. 4.21 Relative changes of power flows. The relative power flow is $\bar{f}_{ij}(t) = \frac{f_{ij}(t)-f_{ij}(0)}{f_{ij}(0)}$, where $f_{ij}(t)$ is the power flow of branch $\{i, j\}$ at time t

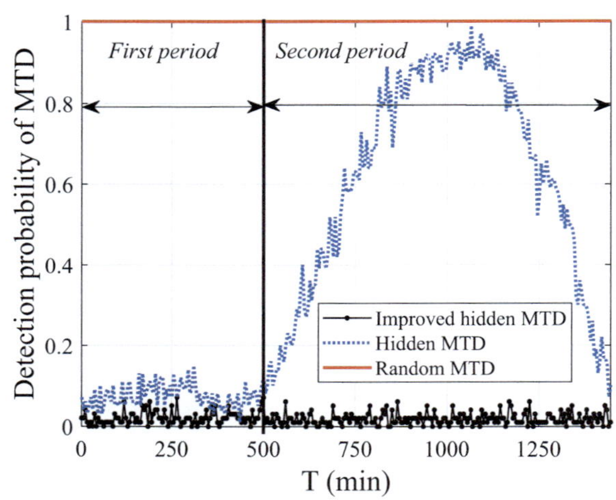

Fig. 4.22 Comparison of the hiddenness among the improved hidden MTD, hidden MTD and random MTD

can see that the detection probability of the improved hidden MTD is around 3% in both periods regardless of the changes of power flows. As for the hidden MTD, the detection probability is small in the first period but increases rapidly in the second period. As for the random MTD without considering the hiddenness, the detection probability approaches 100% in both periods, which means that the random MTD is always detected by the attacker. The simulation results highlight that the improved hidden MTD can be hidable regardless of the changes of power flows, while the hidden MTD and the random MTD cannot.

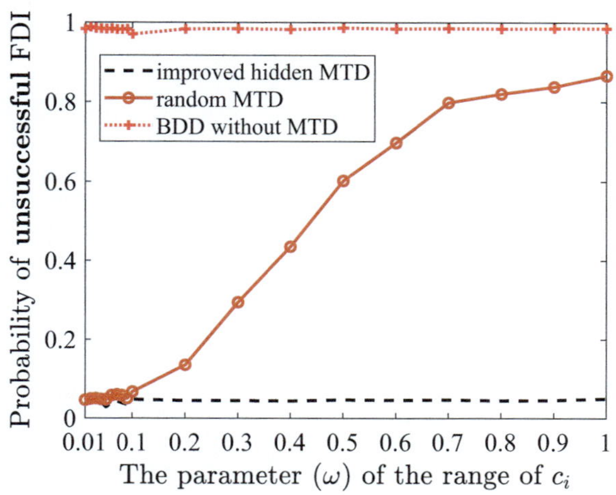

Fig. 4.23 Comparison of MTD's effectiveness between the improved hidden MTD and the random MTD with the IEEE 118-bus power system

IEEE 118-Bus Power System

Moreover, we expand the evaluations of the hiddenness and effectiveness of MTD with the IEEE 118-bus power system. We compare the performance of the improved hidden MTD with the random MTD [44]. The deployment of MTD and MPS are shown in Figs. 4.16, 4.17 and 4.18. Here we use the default setting and data of the IEEE 118-bus power system in MATPOWER.

We evaluate the effectiveness of MTD with respect to the range of the state bias c_i. The attacking pool is generated by uniformly selecting the modified state variables and identically sampling the state bias from a uniform distribution. The perturbation range of MTD is set to be 0.2 for both the improved hidden MTD and the random MTD. The range of c_i is changed from [–0.01, 0.01] to [–1, 1], i.e., $0.01 \leq \omega \leq 1$. For each setting, we construct 1000 attack vectors in the form of $\mathbf{a} = \mathbf{Hc}$. The simulation results are plotted in Fig. 4.23. We can see that the probability of unsuccessful FDI attacks with the improved hidden MTD is always above 90%, even with a small c_i. While with the random MTD, the probability of unsuccessful FDI attacks is always lower than 90%, although it sees an increasing trend with the range of c_i. The result shows that the improved hidden MTD performs better than the random MTD in terms of thwarting FDI attacks. Moreover, the improved hidden MTD has a good scalability to larger power systems. With 300-bus or larger sized power systems, the principle to implement the improved hidden MTD is the same as what we have presented in Sect. 6. It can be proved that this principle is not just designed for small power systems but for power systems of all sizes. Hence here we omit the duplicated simulations for the effectiveness of the improved hidden MTD.

Next, we evaluate the hiddenness of MTD with respect to the branch perturbation range of MTD. To mimic the load change, we add an error within [–10, 10] to each load. The simulation results are plotted in Fig. 4.24. We can see that the detection probabilities of MTD with the improved hidden MTD are around 5% under both

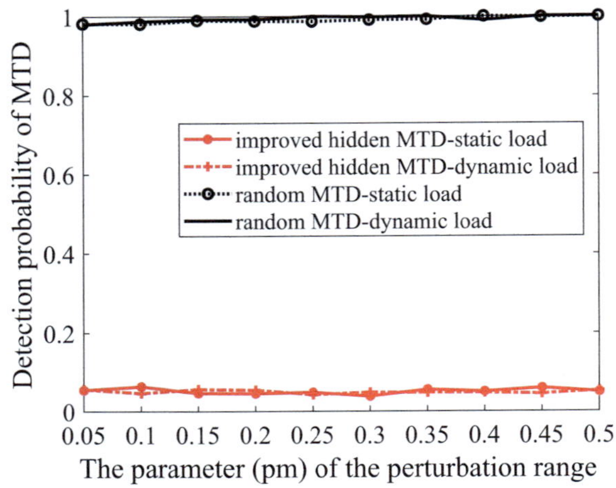

Fig. 4.24 Comparison of MTD's hiddenness between the improved hidden MTD and the random MTD with the IEEE 118-bus power system

the static and dynamic load cases, while they are above 95% with the random MTD under these two cases. It indicates that the improved hidden MTD remains hidable even the system loads are changed.

The above simulation results show that the protection mechanism proposed in this chapter has a good scalability to large-scale power systems.

Achieving Hidable and Effective Moving Target Defense Regardless of Perturbed Branches

Lastly, we verify that we can achieve hidable and effective MTD with the MPS+MTD co-design strategy regardless of the perturbed branches. We use Algorithm 3 to find a spanning tree of the power transmission network of the IEEE 14-bus power system. The spanning tree is formed by branches k_2, k_3, k_5, k_8, k_{10}, k_{13}, k_{14}, k_{15}, k_{16}, k_{17}, k_{18}, k_{19} and k_{20}. The rest branches are k_1, k_4, k_6, k_7, k_9, k_{11} and k_{12}. According to Theorem 4.5, we should protect the measurements z_1, z_4, z_6, z_7, z_9, z_{11} and z_{12}. The deployment of MPS is shown in Fig. 4.25. We assume that the attacker has known the previous susceptances of all branches and the current measurements that are not protected. We realize 4 perturbation schemes for MTD by selecting different sets of perturbed branches. For each scheme, we inject the load data trace from New York State on 09-07-2018. For the hiddenness of MTD, the simulation results are plotted in Fig. 4.26. We can see that the detection probabilities of MTD are 0 in all perturbation schemes, which highlights that the MTD is hidable regardless of the perturbed branches and the changes of power flows. For the effectiveness of MTD, we generate 10000 attack vectors in the form of $\mathbf{a} = \mathbf{Hc}$. The modified state variables (i.e., the non-zero elements in $\mathbf{c}$) are uniformly selected. Besides, the state bias c_i is identically sampled from a uniform distribution $\mathcal{U}(-1, 1)$. We plot the probability of unsuccessful FDI attacks ($\Pr = \frac{\text{\# of unsuccessful attack vectors}}{1000}$) in Fig. 4.26. We can see

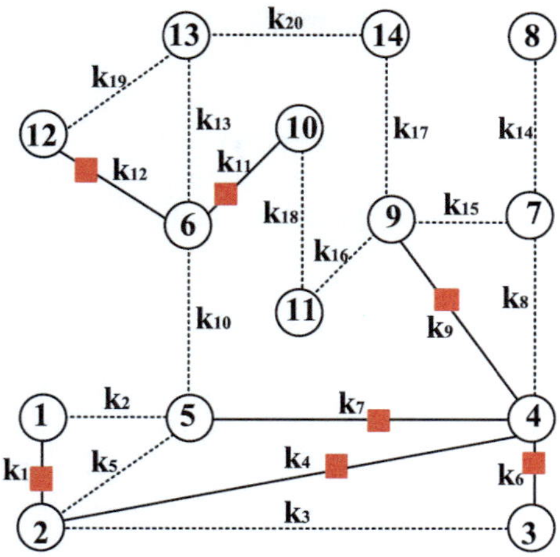

Fig. 4.25 The deployment of MPS for achieving hidable MTD regardless of the perturbed branches

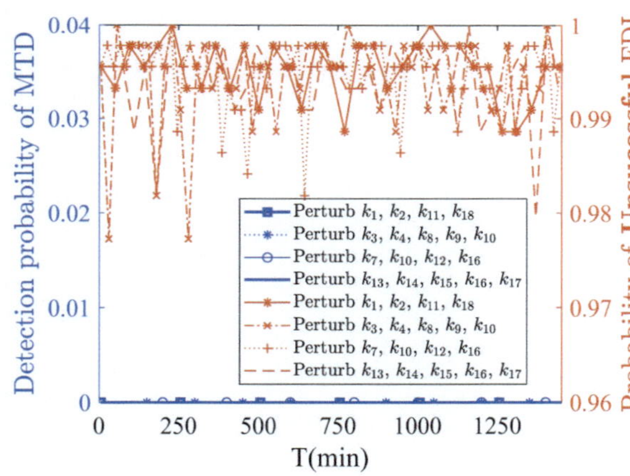

Fig. 4.26 Performance of the MTD+MPS co-design strategy

that Pr is over 90% in all perturbation schemes, which indicates that the improved hidden MTD with the deployment of MPS shown in Fig. 4.25 is effective to thwart FDI attacks.

4.7 Summary

In this chapter, we investigated the conditions needed for achieving a hidable MTD. To begin with, we divided the attackers into partially knowledgeable and fully knowledgeable according to their knowledge of branch parameters and measurements. Then, we constructed an MTD-confriming detector like the BDD based on the previous branch parameters and the measurements after MTD. With the MTD-confriming detector, the attacker is able to detect the activation of MTD. From the attacker's perspective, we proved that an MTD is possibly hidable only if the attacker is partially knowledgeable. With the basic requirement that an MTD must be effective to thwart FDI attacks, we proved that any MTD can be detected by the fully knowledgeable attackers. From the defender's perspective, since it is impossible to know the attacker's capability, we cannot make sure that an MTD is hidable and effective by purely depending on the MTD design. To address this issue, we proposed a MPS+MTD co-design strategy. With a set of measurements being protected, we can achieve a hidable and effective MTD regardless of the changes of power flows, the attacker's knowledge and branch perturbations. Furthermore, we conducted extensive simulations to validate our findings using the IEEE test power systems.

References

1. S. Bi, Y.J. Zhang, Graphical methods for defense against false-data injection attacks on power system state estimation. IEEE Trans. Smart Grid **5**(3), 1216–1227 (2014)
2. R.B. Bobba et al., Detecting false data injection attacks on DC state estimation, in *IEEE/ACM Proceedings of the 1st Conference Workshop Secure Control Systems (CPSWEEK)*, pp. 1–9 (2010)
3. L. Che, X. Liu, Z. Li, Mitigating false data attacks induced overloads using a corrective dispatch scheme. IEEE Trans. Smart Grids **10**(3), 3081–3091 (2019)
4. L. Che, X. Liu, Z. Li, Fast screening out high risk lines under false data injection attack. IEEE Trans. Smart Grids **10**(4), 4003–4014 (2019)
5. G. Dán, H. Sandberg, Stealth attacks and protection schemes for state estimators in power systems, in *Proceedings of the IEEE International Conference on Smart Grid Communications (SmartGridComm)*, pp. 214–219 (2011)
6. K.R. Davis, K.L. Morrow, R. Bobba, E. Heine, Power flow cyber attacks and perturbation-based defense, in *Proceedings of the IEEE Conference on Communications, Control, and Computing Technologies for Smart Grids (SmartGridComm)*, pp. 342–347 (2012)
7. R. Deng, H. Liang, False data injection attacks with limited susceptance information and new countermeasures in smart grid. IEEE Trans. Ind. Informat. **15**(3), 1619–1628 (2019)
8. R. Deng, G. Xiao, R. Lu, H. Liang, A.V. Vasilakos, False data injection on state estimation in power systems-attacks, impacts, and defense: a survey. IEEE Trans. Indust. Inform. **13**(2), 411–423 (2017)
9. R. Deng, G. Xiao, R. Lu, Defending against false data injection attacks on power system state estimation. IEEE Trans. Indust. Inform. **13**(1), 198–207 (2017)
10. R. Deng, P. Zhuang, H. Liang, CCPA: coordinated cyber-physical attacks and countermeausures in smart grid. IEEE Trans. Smart Grid **8**(5), 2420–2430 (2017)
11. D. Divan, H. Johal, Distributed FACTs: a new concept for realizing grid power flow control. IEEE Trans. Power Syst. **22**(6), 2253–2260 (2007)
12. D. Divan, P. Kandula, Distributed power electronics: an enabler for the future grid. CPSS Trans. Power Electron. Appl. **1**(1), 57–65 (2016)
13. M. Esmalifalak, H. Nguyen, R. Zheng, Z. Han, Stealth false data injection using independent component analysis in smart grid, in *Proceedings of the IEEE International Conference on Smart Grid Communications (SmartGridComm)*, pp. 244–248 (2011)
14. A. Giani, E. Bitar, M. Garcia, M. McQueen, P. Khargonekar, K. Poolla, Smart grid data integrity attacks: characterizations and countermeasures, in *Proceedings of the IEEE International Conference on Smart Grid Communications (SmartGridComm)*, pp. 232–237 (2011)
15. A. Giani, E. Bitar, M. Garcia, M. McQueen, P. Khargonekar, K. Poolla, Smart grid data integrity attacks. IEEE Trans. Smart Grid **4**(3), 1244–1253 (2013)

16. C. Gu, J. Panida, M. Mehul, Detecting false data injection attacks in AC state estimation. IEEE Trans. Smart Grid **6**(5), 2476–2483 (2015)
17. V.C. Gungor, D. Sahin, T. Kocak, S. Ergut, C. Buccella, C. Cecati, G.P. Hancke, Smart grid technologies: communication technologies and standards. IEEE Trans. Ind. Inform. **7**(4), 529–539 (2011)
18. Homeland Security, Moving Target Defense (2018). [online]. https://www.dhs.gov/science-and-technology/csd-mtd
19. R.A. Horn, C.R. Johnson, *Matrix Analysis* (Cambridge University Press, New York, 1990)
20. A.G. Illera, J.V. Vidal, Lights off! The darkness of the smart meters (2014), http://youtube.be/Z_y_vjYtAWM
21. T. Jue, R. Tan, X. Guan, T. Liu, Enhanced hidden moving target defense in smart grids. IEEE Trans. Smart Grids 1–1 (2018)
22. T.T. Kim, H.V. Poor, Strategic protection against data injection attacks on power grid. IEEE Trans. Smart Grid **2**(2), 326–333 (2011)
23. S. Lakshminarayana, T.Z. Teng, R. Tan, D.K.Y. Yau, Modeling and detecting false data injection attacks against railway traction power systems. ACM Trans. Cyber Phys. Syst. (TCPS) **2**(4), Article 28 (2018)
24. S. Lakshminaraytana, D.K.Y. Yau, Cost-benefit analysis of moving-target defense in power grids, in *IEEE/IFIP International Conference on Dependable Systems and Networks (DSN)*, pp. 1–12 (2018)
25. R. Langner, Stuxnet: dissecting a cyberwarfare weapon. IEEE Secur. Priv. **9**(3), 49–51 (2011)
26. B. Li, T. Ding, C. Huang, J. Zhao, Y. Yang, Detecting false data injection attacks against power system state estimation with fast go-decomposition (GoDec) approach. IEEE Trans. Ind. Infrom. 1–12 (2018)
27. B. Li, R. Lu, K.-K.R. Choo, W. Wang, S. Luo, On reliability analysis of smart grids under topology attacks: a stochastic petri net approach. ACM Trans. Cyber Phys. Syst. (TCPS) **3**(1), Article 10 (2018)
28. G. Liang, S.R. Weller, J. Zhao, F. Luo, Z.Y. Dong, The 2015 Ukraine blackout: implications for false data injection attacks. IEEE Trans. Power Syst. **32**(4), 3317–3318 (2017)
29. Y. Liu, P. Ning, M.K. Reiter, False data injection attacks against state estimation in electric power grid. IEEE Trans. Inform. Syst. Secur. **14**(1), Article 13 (2011)
30. J. Liu, Y. Xiao, S. Li, W. Liang, C.P. Chen, Cyber security and privacy issues in smart grid. IEEE Commun. Surv. Tutor. **14**(4), 981–997 (2012). Fourth Quarter
31. X. Liu, Z. Li, Local load redistribution attacks in power systems with incomplete network information. IEEE Trans. Smart Grid **5**(4), 1665–1676 (2014)
32. X. Liu, Z. Li, False data attacks against AC state estimation with incomplete network information. IEEE Trans. Smart Grid **8**(5), 2239–2248 (2017)
33. X. Liu, Z. Bao, B. Lu, Z. Li, Modeling of local false data injection attacks with reduced network information. IEEE Trans. Smart Grid **6**(4), 1686–1696 (2015)
34. X. Liu, Z. Li, Z. Li, Optimal protection strategy against false data injection attacks in power systems. IEEE Trans. Smart Grid **8**(4), 1802–1810 (2017)
35. C. Liu, J. Wu, C. Long, D. Kundur, Reactance perturbation for detecting and identifying FDI attacks in power system state estimation. IEEE J. Sel. Top. Signal Process. **12**(4), 763–776 (2018)
36. A. Majumdar, B.C. Pal, Bad data detection in the context of leverage point attacks in modern power. IEEE Trans. Smart Grid **9**(3), 2042–2054 (2018)
37. I. Markwood, Y. Liu, K. Kwiat, C. Kamhoua, Electric grid power flow model camouflage against topology leaking attacks, in *Proceedings of the IEEE Conference on Computer Communications (INFOCOM)*, pp. 1–9 (2017)
38. S. McIntyre, Smart meter security testing framework (2019), https://github.com/securestate/termineter
39. Y. Mo, T.H.-J. Kim, K. Brancik, D. Dickinson et al., Cyber-physical security of a smart grid infrastructure. Proc. IEEE **100**(1), 195–209 (2012)

40. K.L. Morrow, E. Heine, K.M. Rogers, R.B. Bobba, T.J. Overbye, Topology perturbation for detecting malicious data injection, in *Proceedings of the International Conference on System Sciences (CSS)*, pp. 2104–2113 (2012)
41. NESCOR, Electric sector failure scenarios and impact analyses (2014), http://smartgrid.epri.com/doc/NESCOR%20failure%20scenarios%2006-30-14a.pdf
42. NYISO load data. http://www.nyiso.com/public/markets_operations/market_data/load_data/index.jsp
43. S. Pal, B. Sikdar, J.H. Chow, Classification and detection of PMU data manipulation attacks using transmission line parameters. IEEE Trans. Smart Grid **9**(5), 5057–5066 (2018)
44. M.A. Rahman, E. Al-Shaer, R.B. Bobba, Moving target defense for hardening the security of the power system state estimation, in *Proceedings of the 1th ACM Workshop Moving Target Defense (MTD)*, pp. 59–68 (2014)
45. K. Rogers, T. Overbye, Some applications of distributed flexible ac transmission system (d-facts) devices in power systems, in *IEEE 40th North American Power Symposium (NAPS)*, pp. 1–8 (2009)
46. Y. Shoukry, M. Chong, M. Wakaiki, P. Nuzzo et al., SMT-based observer design for cyber-physical systems under sensor attacks. ACM Trans. Cyber Phys. Syst. (TCPS) **2**(1), Article 5 (2018)
47. M. Stephen, P. Dmitry, M. Sergei, D. Adam, M. Patrick, Multi-vendor penetration testing in the advanced metering infrastructure, in *Proceedings of the ACM 26th Annual Computer Security Applications Conference (ACSAC)*, pp. 107–116 (2010)
48. J. Yao, P. Venkitasubramaniam, S. Kishore, L.V. Snyder, R.S. Blum, Network topology risk assessment of stealthy cyber attacks on advanced metering infrastructure networks, in *IEEE 51st Annual Conference on Information Sciences and Systems (CISS)*, pp. 1–6 (2017)
49. S.Z. Yong, M. Zhu, E. Frazzoli, Switching and data injection attacks on stochastic cyber-physical systems: modeling, resilient estimation, and attack mitigation. ACM Trans. Cyber Phys. Syst. (TCPS) **2**(2), Article 9 (2018)
50. Z. Zhang, R. Deng, D.K.Y. Yau, P. Cheng, J. Chen, Analysis of moving target defense against false data injection attacks on power grid. IEEE Trans. Inf. Forensics Secur. (2019)
51. Z. Zhang, R. Deng, D.K.Y. Yau, P. Cheng, J. Chen, On effectiveness of detecting FDI attacks on power grid using moving target defense, in *IEEE PES Innovative Smart Grid Technologies Conference (ISGT)*, pp. 1–5 (2019)
52. Z. Zhang, Y. Qin, J. Wang, H. Li, R. Deng, Detecting the one-shot dummy attack on the power industrial control processes with an unsupervised data-driven approach. IEEE/CAA J. Autom. Sin. **10**(2), 550–553 (2023)
53. G. Zhang, Z. Wang, X. Fan, P. Li, D. Gao, Z. Zhang, Q. Wan, Y. Zhang, Self-rectifying memristor-based reservoir computing for real-time intrusion detection in cybersecurity. Nano Lett. **24**(49), 15707–15715 (2024)
54. R.D. Zimmerman, C.E. Murillo-Sanchez, R.J. Thomas, MATPOWER: steady-state operations, planning, and analysis tools for power systems research and education. IEEE Trans. Power Syst. **26**(1), 12–19 (2011)

Chapter 5
Explicit Analysis of Moving Target Defense in the Smart Grid

5.1 System Model and Problem Formulation

In this section, we introduce the AC power system model, SE model, BDD model, threat model and reactance perturbation model. Throughout this chapter, we utilize $(\cdot) \in \mathbb{C}$ to denote the complex number and $(\cdot) \in \mathbb{R}$ to indicate the real number. Moreover, $(\cdot)^{\mathrm{H}}$ signifies the conjugate, $(\cdot)^{\mathrm{T}}$ denotes the transpose, and $||\cdot||_p$ represents the p-norm. The imaginary unit of complex number is denoted by $\jmath$. Inequalities between vectors are compared by elements. The terms "bus" and "node", "branch" and "power line" are used interchangeably.

5.1.1 The Smart Grid and AC Models

(1) AC Power Flow: We focus on a symmetric and balanced power system, for which a single-phase model is commonly utilized. The model is composed of N nodes and L power lines. The set of nodes is denoted by $\mathcal{N} \triangleq \{0, \mathcal{N}^{\mathrm{PV}}, \mathcal{N}^{\mathrm{PQ}}\}$, where 0 signifies the reference bus, $\mathcal{N}^{\mathrm{PV}}$ represents the set of PV buses, and $\mathcal{N}^{\mathrm{PQ}}$ implies the set of PQ buses. The set of power lines is denoted by $\mathcal{L} \triangleq \{(j,k)\}$, and each branch (j,k) connects buses j and k. The set of nodes connecting to node $j \in \mathcal{N}$ is denoted by $\mathcal{N}_j$. In this chapter, we only consider the steady states of the power system, where all voltages are sinusoidal signals at the same frequency. The voltage and current signals can therefore be characterized as $(\cdot) = |\cdot|e^{\jmath\angle\cdot}$, where $|\cdot|$ is the root-mean-square value of the signal and $\angle\cdot$ is the phase angle with respect to the reference bus.

The nodal voltages are taken to form the system state vector $x \triangleq [\theta_{\mathcal{N}^{\mathrm{PV}}}; \theta_{\mathcal{N}^{\mathrm{PQ}}}; V_{\mathcal{N}^{\mathrm{PQ}}}] \in \mathbb{R}^n$, where vectors $\theta_{\mathcal{N}^{\mathrm{PV}}}$ and $\theta_{\mathcal{N}^{\mathrm{PQ}}}$ contain the phase angles of PV and PQ buses, respectively, and vector $V_{\mathcal{N}^{\mathrm{PQ}}}$ includes the voltage magnitudes of PQ buses. Here $n = 2(N-1) - N^{\mathrm{PV}}$ with N^{PV} being the number of PV buses.

R. Deng et al., *Moving Target Defense in the Smart Grid*, Wireless Networks,
https://doi.org/10.1007/978-3-031-92378-4_5

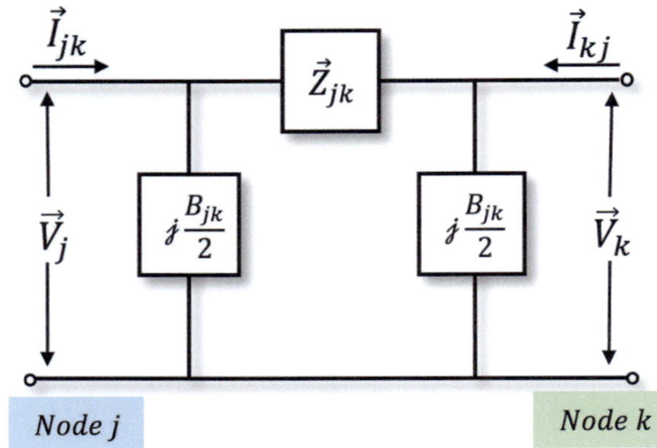

Fig. 5.1 The π branch model

Once x is given, all power flow and power injection measurements can be determined. We usually have $m > n$ measurements to ensure that the state vector is fully observable from the meter measurements [23]. The mathematic relation between the measurement vector $z \in \mathbb{R}^m$ and x is denoted by

$$z \triangleq h(x) + e, \tag{5.1}$$

where vector $h(\cdot)$ contains the measurement functions and vector e is composed of measurement noises following normal distributions. We consider the standard π branch model as illustrated in Fig. 5.1, where $\mathbf{Z}_{jk} \triangleq R_{jk} + jX_{jk}$ denotes the series impedance and B_{jk} signifies the total charging susceptance. The series admittance is denoted by $\mathbf{Y}_{jk} \triangleq (\mathbf{Z}_{jk})^{-1}$.

In practice, the measurements received from remote terminal units include active/reactive power flows, real/reactive power injections, and voltage magnitudes, i.e., $z \triangleq [z_1; z_2; z_3]$. Here, vector $z_1 \triangleq [P_{\mathcal{L}}^{\text{fwd}}; Q_{\mathcal{L}}^{\text{fwd}}; P_{\mathcal{L}}^{\text{rvs}}; Q_{\mathcal{L}}^{\text{rvs}}]$ contains active and reactive power flows of branches $\mathcal{L}$ in both forward and reverse directions; vector $z_2 \triangleq [P_{\mathcal{N}}; Q_{\mathcal{N}}]$ includes active and reactive power injections of buses $\mathcal{N}$; and vector $z_3 \triangleq V_{\mathcal{N}^{\text{PQ}}}$ is composed of voltage magnitudes of PQ buses $\mathcal{N}^{\text{PQ}}$. The measurement function of power flow $(j,k) \in \mathcal{L}$ is

$$h_{jk}^P(x) + \mathrm{j} h_{jk}^Q(x) \triangleq \mathbf{V}_j \left[\mathbf{Y}_{jk}(\mathbf{V}_j - \mathbf{V}_k) + \mathrm{j}\frac{B_{jk}}{2}\mathbf{V}_j \right]^{\mathrm{H}}.$$

The measurement function of power injection at bus $j \in \mathcal{N}$ is

$$h_j^P(x) + \mathrm{j} h_j^Q(x) \triangleq \sum_{i \in \mathcal{N}_j} \mathbf{V}_j \left[\mathbf{Y}_{ji}(\mathbf{V}_j - \mathbf{V}_i) + \mathrm{j}\frac{B_{ji}}{2}\mathbf{V}_j \right]^{\mathrm{H}}.$$

The measurement function of voltage magnitude at bus $j \in \mathcal{N}^{\mathrm{PQ}}$ is

$$h_j^V(x) \triangleq |\mathbf{V}_j|.$$

For the large-scale power system with numerous measurement points, it is impractical to telemeter all points through real-time measuring devices such as phasor measurement units, intelligent electronic devices, and advanced metering infrastructure systems [24, 25]. To guarantee the observability from the measurements to the state vector, the pseudo measurements produced by the control center through historical customer load profiles, are usually utilized as the pseudo power injection measurements [26].

(2) AC State Estimation: SE is to obtain the estimate of x, denoted by x^*, that is the best fit of z following (5.1). The NWLS problem is usually formulated to find x^* [27], i.e.,

$$x^* \triangleq \arg\min_x J(x) = \arg\min_x r^{\mathrm{T}}(x) W r(x), \tag{5.2}$$

where $r(x) \triangleq z - h(x)$ calculates the residual vector and the diagonal weight matrix $W \triangleq \mathrm{diag}([\delta_1^{-2}, \ldots, \delta_m^{-2}])$ with δ_i being the standard deviation of i-th measurement noise. The common method to solve (5.2) is the iterative Gauss-Newton method [28], which is based on a linear approximation to the objective function. Specifically, for the i-th iteration at $x = x_{[i]}$, when the step size $||\Delta_{[i]}||_2$ is small, the Taylor expansion of the residual function $r(x_{[i]} + \Delta_{[i]})$ can be approximated as

$$\begin{aligned} r(x_{[i]} + \Delta_{[i]}) &= r(x_{[i]}) - H_{x_{[i]}} \Delta_{[i]} + O(||\Delta_{[i]}||_2^2) \\ &\approx l(\Delta_{[i]}) \triangleq r(x_{[i]}) - H_{x_{[i]}} \Delta_{[i]}, \end{aligned} \tag{5.3}$$

where $H_{x_{[i]}}$ denotes the Jacobian matrix of $h(x)$ at $x = x_{[i]}$. Substituting (5.3) into the objective function $J(x)$, we obtain

$$\begin{aligned} J(x_{[i]} + \Delta_{[i]}) &\approx L(\Delta_{[i]}) \triangleq \frac{1}{2} l^{\mathrm{T}}(\Delta_{[i]}) W l(\Delta_{[i]}) \\ &= \frac{1}{2} J(x_{[i]}) - \Delta_{[i]}^{\mathrm{T}} H_{x_{[i]}}^{\mathrm{T}} W r(x_{[i]}) + \frac{1}{2} \Delta_{[i]}^{\mathrm{T}} H_{x_{[i]}}^{\mathrm{T}} W H_{x_{[i]}} \Delta_{[i]}, \end{aligned}$$

under which the gradient and Hessian of $L(\Delta_{[i]})$ can be calculated as

$$\begin{aligned} L^{'}(\Delta_{[i]}) &= -H_{x_{[i]}}^{\mathrm{T}} W r(x_{[i]}) + H_{x_{[i]}}^{\mathrm{T}} W H_{x_{[i]}} \Delta_{[i]}, \\ L^{''}(\Delta_{[i]}) &= H_{x_{[i]}}^{\mathrm{T}} W H_{x_{[i]}}. \end{aligned}$$

If matrix $H_{x_{[i]}}$ is of full column rank, then $L^{''}(\Delta_{[i]})$ will be positive definite. This indicates that $L(\Delta_{[i]})$ has a unique minimizer (by solving $L^{'}(\Delta_{[i]}) = 0$)

$$\Delta_{[i]} \triangleq \left[H_{x_{[i]}}^{\mathrm{T}} W H_{x_{[i]}}\right]^{-1} H_{x_{[i]}}^{\mathrm{T}} W r(x_{[i]}), \tag{5.4}$$

which is utilized to update $x_{[i]}$ as

$$x_{[i+1]} \triangleq x_{[i]} + \Delta_{[i]}.$$

The iteration is repeated until $\left\|\Delta_{[i]}\right\|_{\infty}$ and $\left|J\left(x_{[i+1]}\right) - J\left(x_{[i]}\right)\right|$ are small enough [29]. Indeed, the accuracy of the update rule (5.4) relies heavily on the approximation error of the Taylor expansion (5.3). In particular, the Taylor expansion will be more accurate if the step size $||\Delta_{[i]}||$ is smaller, which is mainly determined by the residual vector $r(x_{[i]})$ according to (5.4). This implies that if the initial value $x_{[0]}$ is selected close to the optimal value, then the accuracy of the update rule (5.4) will be increased and can promote the convergence.

Due to stochastic meter failures and malicious cyber attacks, there may exist bad data in numerous measurements. Based on the result of SE, the system operator employs the residual-based BDD to perceive bad data, and to identify and eliminate it if possible [30]. In particular, the measurement residual is obtained by comparing the actual measurement z with the estimated measurement $\hat{z}$, i.e.,

$$r \triangleq ||r(x^*)||_2 = ||z - \hat{z}||_2 = ||z - h(x^*)||_2.$$

When measurement noises follow normal distributions, r^2 will follow the chi-square distribution with $(m - n)$ freedom, i.e., χ^2_{m-n}, in the normal case. Through the hypothesis test, a predetermined threshold τ can be given with a significance level α [30]. If the residual r is larger than τ, i.e., $r > \tau$, then an abnormal alarm will be triggered by the BDD, whose false alarm rate is equal to the significance level; otherwise, z is taken as normal and the hypothesis is accepted.

(3) Threat Model: This chapter considers the worst case where the attacker has the following capabilities:

- The attacker can eavesdrop and tamper with measurements through spoofed communication signals, intruded shared communications, or spoofed field devices [31].
- The attacker can infer the network topology and branch parameters of power system through topology leaking attacks [15] and subspace attacks [16], respectively, which typically requires the order of a few hours.
- The attacker can approximate the system state vector based on power flow or power injection measurements without too many efforts as shown in [4].

With the above capabilities, the attacker is able to launch the FDI attack that can bypass the BDD. The adopted actions are summarized as follows: (1) inferring the measurement function $h(\cdot)$; (2) approximating the system state x^* as x^{appr}; (3) constructing the attack vector as

$$a \triangleq h(x^{\text{appr}} + c) - h(x^{\text{appr}}), \tag{5.5}$$

where $c \in \mathbb{R}^n$ denotes the bias vector that the attacker intends to inject into the state vector; (4) tampering with z using a. In the absence of MTD, it is evident that the measurement residual under the FDI attack is

$$\begin{aligned} r_a &\triangleq ||z + a - h(x_a^*)||_2 \\ &= ||z + h(x^{\text{appr}} + c) - h(x^{\text{appr}}) - h(x_a^*)||_2 \\ &\approx ||z - h(x^{\text{appr}})||_2 \approx ||z - h(x^*)||_2 \approx r, \end{aligned}$$

due to the facts that $x^{\text{appr}} \approx x^*$ and $x_a^* \approx x^* + c \approx x^{\text{appr}} + c$, where x_a^* denotes the estimated state vector under the FDI attack. The falsified system states by the FDI attack can make the system operator unconscious of the current system states, and thus will invalidate the application functions after SE such as the contingency analysis, emergency control, restorative control, and load forecasting [30], which may cause catastrophic failures to the power system.

(4) Reactance Perturbation Model: For branch $(j, k) \in \mathcal{L}$ equipped with the D-FACTS device, the reactance can be arbitrarily perturbed within

$$X_{jk} + \Delta \underline{X}_{jk} \leq X_{jk}^{\text{MTD}} \leq X_{jk} + \Delta \overline{X}_{jk}, \tag{5.6}$$

where X_{jk}^{MTD} denotes the branch reactance after MTD, and $\Delta \underline{X}_{jk}$ and $\Delta \overline{X}_{jk}$ signify the lower and upper bounds for the perturbation command ΔX_{jk}, respectively. Let vector b contain parameters R_{jk}, X_{jk}, and B_{jk} of all branches, and let vector Δb contain the perturbation commands. We have

$$\Delta \underline{b} \leq \Delta b \leq \Delta \overline{b}, \tag{5.7}$$

where the elements in $\Delta \underline{b}$ and $\Delta \overline{b}$ are set as zeros when the elements correspond to series resistances, charging susceptances, or the branches equipped with no D-FACTS device. To defend against the FDI attack, the system operator in the control center will command the remote D-FACTS devices to proactively perturb the branch reactances such that the previously inferred $h(\cdot)$ can be antiquated. Here, the perturbation commands are transmitted through encrypted secure channels to guarantee the confidentiality and integrity. Since the attacker requires *several hours* to complete the inference process for the network topology and branch parameters (i.e., $h(\cdot)$) [16], the perturbation commands should be updated quickly than that (e.g., hourly) to invalidate the inference process.

5.1.2 Problems of Interest

In this section, we quantify the effectiveness and hiddenness of MTD based on measurement residuals and introduce our problems of interest. With some abuse of notations, we use symbols without subscript such as $h(\cdot), x, b, z$ to denote the quantities before MTD, and symbols with subscript $(\cdot)_{\text{MTD}}$ to signify those after MTD. For clarity, we provide a graphical illustration in Fig. 5.2 for the sequence of actions adopted by the attacker and the system operator.

(1) The Hiddenness of MTD: Before launching FDI attacks, the attacker will apply BDD to eavesdropped measurements to judge the consistency between the inferred model knowledge and the current system model. Here, we consider the worst case where the attacker can obtain the same measurements as the system operator, i.e., z_{MTD}, to evaluate the hiddenness of MTD. The estimated system state is obtained by solving

$$x_{\text{att}}^* \triangleq \arg\min_x J_{\text{att}}(x) = \arg\min_x r_{\text{att}}^{\text{T}}(x) W r_{\text{att}}(x),$$

where $r_{\text{att}}(x) \triangleq z_{\text{MTD}} - h(x)$. We note that the power flow and voltage magnitude measurements of z_{MTD} can be different from those of z. The hiddenness of MTD is quantified as

$$r_{\text{att}} \triangleq \left\| r_{\text{att}}(x_{\text{att}}^*) \right\|_2 . \tag{5.8}$$

Due to the inconsistency between $h(\cdot)$ and $h_{\text{MTD}}(\cdot)$, z_{MTD} may not be totally explained by $h(\cdot)$ in the absence of measurement noises, i.e., it may be difficult to find a x such that $h(x)$ approaches z_{MTD} infinitely. Hence, r_{att} would be nontrivial if the perturbation commands are not properly designed. From the point view of the attacker, the smaller r_{att} means the stronger hiddenness. Only if the MTD strategy is hidden from the attacker, then the FDI attack will be launched. Otherwise, the attacker will restart the inference process, and the FDI attack will not be launched.

(2) The Effectiveness of MTD: In the presence of MTD, the corrupted measurement vector is $z_{\text{MTD}}^a = z_{\text{MTD}} + a_{\text{MTD}}$, where the construction of attack vector

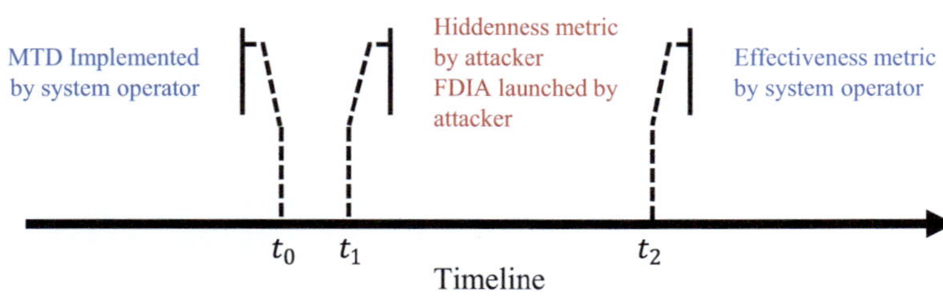

Fig. 5.2 Graphical illustration for the sequence of actions adopted by the attacker and the system operator, where $t_0 \leq t_1 \leq t_2$

$a_{\text{MTD}} = h(x_{\text{MTD}}^{\text{appr}} + c) - h(x_{\text{MTD}}^{\text{appr}})$ is similar to (5.5). Here, $x_{\text{MTD}}^{\text{appr}}$ denotes the attacker's approximation of the estimated state vector by the system operator after MTD, i.e., x_{MTD}^*, such that $h_{\text{MTD}}(x_{\text{MTD}}^*) = z_{\text{MTD}} \approx h_{\text{MTD}}(x_{\text{MTD}}^{\text{appr}})$.[1] Due to the inconsistency between $h(\cdot)$ and $h_{\text{MTD}}(\cdot)$ caused by MTD, z_{MTD}^a would deviate from the expected vector $h_{\text{MTD}}(x_{\text{MTD}}^{\text{appr}} + c)$. Thus, it may be difficult to find a x such that $h_{\text{MTD}}(x)$ approaches z_{MTD}^a infinitely in the absence of measurement noises, i.e., z_{MTD}^a may not be totally explained by the output of $h_{\text{MTD}}(\cdot)$, under which the FDI attack may be detected by the BDD. Specifically, the estimated state vector is obtained by solving

$$x_{\text{sys}}^* \triangleq \arg\min_x J_{\text{sys}}(x) = \arg\min_x r_{\text{sys}}^{\text{T}}(x) W r_{\text{sys}}(x),$$

where $r_{\text{sys}}(x) \triangleq z_{\text{MTD}}^a - h_{\text{MTD}}(x)$. The effectiveness of MTD is quantified as

$$r_{\text{sys}} \triangleq \left\| r_{\text{sys}}(x_{\text{sys}}^*) \right\|_2. \tag{5.9}$$

Obviously, r_{sys} needs to be large enough to expose the FDI attack to BDD, and thus the larger r_{sys} means the stronger effectiveness.

(3) Problems of Interest: In this chapter, based on r_{att} and r_{sys}, we attempt to conduct an explicit analysis on the effectiveness and hiddenness and design EXR-MTD in AC power systems. We focus on the noiseless case where the residual is deterministic once the reactance perturbation and the FDI attack are known, and simulations are conducted to validate the applicability of the proposed approach in the presence of measurement noises. Here the challenges lie in the implicit expressions of r_{sys} and r_{att} and the nontrivial trade-off between effectiveness and hiddenness [9]. Hence, our problems of interest include: (1) deriving explicit approximations for r_{sys} and r_{att}; (2) designing EXR-MTD such that r_{sys} is maximized and small r_{att} can be guaranteed.

5.2 Explicit Analysis of Effectiveness and Hiddenness

In this section, we derive explicit approximations for r_{att} and r_{sys} based on the sensitivity analysis, which discusses "how" and "how much" variations in the parameters of an optimization problem will change the optimal objective function value and the optimal solution [32]. Moreover, the accuracy of the derived approximations is validated through numerical results.

---

[1] The impact of the reactance perturbation on the attacker's approximation of the estimated state vector is not considered.

5.2.1 Sensitivity Analysis of Nonlinear Weighted Least Squares Problem

In this subsection, we apply sensitivity analysis to the general NWLS problem (5.2) and derive two essential sensitivities, with which r_{att} and r_{sys} can be approximated in a real-time manner. For brevity, let $\Theta^* \triangleq (x^*, z, b)$ denote the aggregated local optimal point. The first-order Karush-Kuhn-Tucker (KKT) condition for solving the NWLS problem (5.2) is established as

$$\nabla_x J(\Theta^*) = 0. \tag{5.10}$$

The sensitivity is obtained by perturbing the system state x^* and corresponding parameters z, b around Θ^* such that the KKT condition (5.10) still holds. Hence, by differentiating (5.2) and (5.10), we have

$$\begin{bmatrix} J_x^{\mathrm{T}} & J_z^{\mathrm{T}} & J_b^{\mathrm{T}} & -1 \\ J_{xx} & J_{xz} & J_{xb} & 0 \end{bmatrix} \begin{bmatrix} dx \\ dz \\ db \\ dJ \end{bmatrix} = 0, \tag{5.11}$$

where

$$J_x \triangleq \nabla_x J(\Theta^*),\ J_z \triangleq \nabla_z J(\Theta^*),\ J_b \triangleq \nabla_b J(\Theta^*),$$
$$J_{xx} \triangleq \nabla_{xx} J(\Theta^*),\ J_{xz} \triangleq \nabla_{xz} J(\Theta^*),\ J_{xb} \triangleq \nabla_{xb} J(\Theta^*).$$

Proposition 5.1 *When the objective function value $J(\Theta^*)$ approaches zero infinitely, the sensitivities $\frac{\partial x}{\partial b}$ and $\frac{\partial r}{\partial z}$ at point Θ^* are obtained as*

$$\frac{\partial x}{\partial b}|_{\Theta^*} \triangleq -\left[(H_x^*)^{\mathrm{T}} H_x^*\right]^{-1} (H_x^*)^{\mathrm{T}} H_b^*, \tag{5.12}$$
$$\frac{\partial r}{\partial z}|_{\Theta^*} \triangleq I - H_x^*\left[(H_x^*)^{\mathrm{T}} H_x^*\right]^{-1} (H_x^*)^{\mathrm{T}}, \tag{5.13}$$

where H_x^* denotes the Jacobian matrix at $x = x^*$ and $H_b^* \triangleq \partial h(x^*)/\partial b$.

Proof ased on (5.11), we have

$$\boldsymbol{J_{xx} dx} = -\boldsymbol{J_{xz} dz} - \boldsymbol{J_{xb} db}, \tag{5.14a}$$
$$\boldsymbol{J_x^{\mathrm{T}} dx} - dJ = -\boldsymbol{J_z^{\mathrm{T}} dz} - \boldsymbol{J_b^{\mathrm{T}} db}. \tag{5.14b}$$

With (5.14a), we can derive the sensitivities of system states $\boldsymbol{x}$ with respect to z and $\boldsymbol{b}$ as follows

$$\frac{\partial \boldsymbol{x}}{\partial z} = -J_{\boldsymbol{xx}}^{-1} J_{\boldsymbol{x}z}, \tag{5.15}$$

$$\frac{\partial \boldsymbol{x}}{\partial \boldsymbol{b}} = -J_{\boldsymbol{xx}}^{-1} J_{\boldsymbol{xb}}. \tag{5.16}$$

It follows from (5.14b) that

$$\frac{\partial J}{\partial z} = (\frac{\partial \boldsymbol{x}}{\partial z})^{\mathrm{T}} J_x + J_z. \tag{5.17}$$

Substituting (5.15) into (5.17), we obtain

$$\frac{\partial J}{\partial z} = J_z - (J_{\boldsymbol{xx}}^{-1} J_{xz})^{\mathrm{T}} J_x. \tag{5.18}$$

Moreover, the expressions of $J_{\boldsymbol{x}}$, J_z, and $J_{\boldsymbol{x}z}$ can be derived from (5.2) as

$$\begin{aligned} J_x &= -(H_{\boldsymbol{x}}^*)^{\mathrm{T}}\big[z - \boldsymbol{h}(\boldsymbol{x}^*)\big], \\ J_z &= z - \boldsymbol{h}(\boldsymbol{x}^*), \quad J_{xz} = -(H_{\boldsymbol{x}}^*)^{\mathrm{T}}. \end{aligned} \tag{5.19}$$

Substituting (5.19) into (5.18), we have

$$\frac{\partial J}{\partial z} = \big[I - H_{\boldsymbol{x}}^*(J_{\boldsymbol{xx}}^{-1})^{\mathrm{T}}(H_{\boldsymbol{x}}^*)^{\mathrm{T}}\big]\boldsymbol{r}(\boldsymbol{x}^*). \tag{5.20}$$

With $\boldsymbol{r}(\boldsymbol{x}^*) = z - \boldsymbol{h}(\boldsymbol{x}^*)$, we can derive $\frac{\partial J}{\partial z} = (\frac{\partial \boldsymbol{r}}{\partial z})^{\mathrm{T}}\boldsymbol{r}$. Hence, according to (5.20), we have

$$\frac{\partial \boldsymbol{r}}{\partial z} = I - H_{\boldsymbol{x}}^* J_{\boldsymbol{xx}}^{-1}(H_{\boldsymbol{x}}^*)^{\mathrm{T}}. \tag{5.21}$$

Nevertheless, the sensitivities (5.16) and (5.21) are still implicit due to the existence of $J_{\boldsymbol{xx}}$ and $J_{\boldsymbol{x}t}$. To this end, we expand $J_{\boldsymbol{xx}}$ and $J_{\boldsymbol{x}t}$ as follows

$$J_{\boldsymbol{xx}} = (H_{\boldsymbol{x}}^*)^{\mathrm{T}} H_{\boldsymbol{x}}^* - \tilde{J}_{\boldsymbol{xx}}, \tag{5.22}$$

$$J_{\boldsymbol{x}t} = (H_{\boldsymbol{x}}^*)^{\mathrm{T}} H_{\boldsymbol{b}}^* - \tilde{J}_{\boldsymbol{x}t}, \tag{5.23}$$

where $\tilde{J}_{\boldsymbol{xx}} \in \mathbb{R}^{n*n}$ and its (i, j)th entry is $(\tilde{J}_{\boldsymbol{xx}})_{ij} = \big[\frac{\partial \boldsymbol{h}^2(\boldsymbol{x}^*)}{\partial x_i \partial x_j}\big]^{\mathrm{T}}\boldsymbol{r}(\boldsymbol{x}^*)$, and similarly $\tilde{J}_{\boldsymbol{x}t} \in \mathbb{R}^{n*3L}$ and its (i, j)th entry is $(\tilde{J}_{\boldsymbol{x}t})_{ij} = \big[\frac{\partial \boldsymbol{h}^2(\boldsymbol{x}^*)}{\partial x_i \partial t_j}\big]^{\mathrm{T}}\boldsymbol{r}(\boldsymbol{x}^*)$. If the optimal objective value $J^* = \boldsymbol{r}(\boldsymbol{x}^*)^{\mathrm{T}}\boldsymbol{r}(\boldsymbol{x}^*) \to 0$, then $\boldsymbol{r}(\boldsymbol{x}^*) \to \mathbf{0}$ also holds, under which we obtain

$J_{xx} \to (H_x^*)^{\mathrm{T}} H_x^*$ and $J_{xt} \to (H_x^*)^{\mathrm{T}} H_b^*$. Hence, Proposition 5.1 is proved by substituting (5.22) into (5.16) and (5.21).

It is noted that $J(\Theta^*)$ can approach zero infinitely in the absence of measurement noises, and thus (5.12) and (5.13) are directly adopted hereafter. Moreover, the correctness of the derived sensitivities can be easily verified by linearizing the nonlinear measurement function (5.1) around x^* as $\Delta z \triangleq z - z^* = H_x^*(x - x^*) + e = H_x^* \Delta x + e$, and then according to the simplified DC model the alteration of residual $\Delta r = \Delta z - H_x^* \Delta \hat{x}$ can be calculated as $\Delta r \triangleq \Big[I - H_x^*\big[(H_x^*)^{\mathrm{T}} H_x^*\big]^{-1}(H_x^*)^{\mathrm{T}}\Big]\Delta z$, i.e., (5.13).

5.2.2 Approximation of Residuals for Effectiveness and Hiddenness

In this subsection, based on (5.12) and (5.13), explicit approximations of r_{att} and r_{sys} are derived.

(1) Approximation of r_{att}: According to (5.8), the nontrivial r_{att} is caused by the measurement variations before and after MTD, i.e., $\Delta z_{\mathrm{att}} \triangleq h_{\mathrm{MTD}}(x^*_{\mathrm{MTD}}) - h(x^*)$, where x^*_{MTD} denotes the solution to (5.2) with measurement z_{MTD} and measurement function $h_{\mathrm{MTD}}(\cdot)$. Hence, via (5.13), r_{att} is approximated around Θ^* as

$$r_{\mathrm{att}} \approx \left\| \frac{\partial r}{\partial z}|_{\Theta^*} \times \Delta z_{\mathrm{att}} \right\|_2 . \tag{5.24}$$

Here, the computation of precise Δz_{att} requires to requires to solve the AC power flow problem [33], which is typically based on the iterative Newton-Raphson method and is time-consuming. We choose to approximate Δz_{att} utilizing Δb and $\Delta x^* \triangleq x^*_{\mathrm{MTD}} - x^*$, i.e.,

$$\Delta z_{\mathrm{att}} \approx H_b^* \Delta b + H_x^* \Delta x^* \approx \Delta z_{\mathrm{att}}^{\mathrm{appr}} \triangleq \left(H_b^* + H_x^* \times \frac{\partial x}{\partial b}|_{\Theta^*}\right)\Delta b. \tag{5.25}$$

Integrating (5.24) with (5.25), we obtain the explicit approximation of r_{att} as

$$r_{\mathrm{att}} \approx r_{\mathrm{att}}^{\mathrm{appr}} \triangleq \left\|\Delta z_{\mathrm{att}}^{\mathrm{appr}}\right\|_2, \tag{5.26}$$

which implies that the quantified metric for hiddenness is equivalent to the approximated power flow variations before and after MTD, which matches the PFI condition for the hidden MTD in the DC model [14].

(2) Approximation of r_{sys}: In the presence of MTD, the solution to (5.2) with measurement z^a_{MTD} and measurement function $h_{\mathrm{MTD}}(\cdot)$ may slightly deviate from $x^*_{\mathrm{MTD}} + c$ due to the bias $\Delta z_{\mathrm{sys}} \triangleq z^a_{\mathrm{MTD}} - h_{\mathrm{MTD}}(x^*_{\mathrm{MTD}} + c)$, which can result in

nontrivial r_{sys}. Hence, r_{sys} is approximated around the shifted local optimal point $\Theta^*_{\text{MTDa}} \triangleq \big(x^*_{\text{MTD}} + c, h_{\text{MTD}}(x^*_{\text{MTD}} + c), b_{\text{MTD}}\big)$ by the attacker as

$$r_{\text{sys}} \approx \left\| \frac{\partial r}{\partial z}|_{\Theta^*_{\text{MTDa}}} \times \Delta z_{\text{sys}} \right\|_2, \tag{5.27}$$

where Δz_{sys} is expanded as

$$\begin{aligned}\Delta z_{\text{sys}} &= z_{\text{MTD}} + a_{\text{MTD}} - h_{\text{MTD}}(x^*_{\text{MTD}} + c) \\ &\approx -h_{\text{MTD}}(x^*_{\text{MTD}} + c) + h(x^*_{\text{MTD}} + c) + h_{\text{MTD}}(x^*_{\text{MTD}}) - h(x^*_{\text{MTD}}),\end{aligned} \tag{5.28}$$

owing to the fact that x^*_{MTD} can be accurately approximated by the attacker as $x^{\text{appr}}_{\text{MTD}}$ without too many efforts [4]. According to (5.28), there exist two directions for the further approximation of Δz_{sys}, i.e., utilizing the reactance perturbation Δb or the induced state bias c. Specifically, based on Δb, we have

$$\Delta z_{\text{sys}} \approx \Delta z^{\text{apprb}}_{\text{sys}} \triangleq (H_b^{\text{MTD}*} - H_b^{\text{MTDc}*}) \times \Delta b, \tag{5.29}$$

where

$$H_b^{\text{MTD}*} \triangleq \frac{\partial h(x^*_{\text{MTD}})}{\partial b}, \quad H_b^{\text{MTDc}*} \triangleq \frac{\partial h(x^*_{\text{MTD}} + c)}{\partial b}.$$

Based on c, we have

$$\Delta z_{\text{sys}} \approx \Delta z^{\text{apprx}}_{\text{sys}} \triangleq (H^*_{x_{\text{MTD}}} - H^{\text{MTD}*}_{x_{\text{MTD}}}) \times c, \tag{5.30}$$

where

$$H^*_{x_{\text{MTD}}} \triangleq \frac{\partial h(x^*_{\text{MTD}})}{\partial x}, \quad H^{\text{MTD}*}_{x_{\text{MTD}}} \triangleq \frac{\partial h_{\text{MTD}}(x^*_{\text{MTD}})}{\partial x}.$$

Substituting (5.29)–(5.30) into (5.27), the two explicit approximations of r_{sys} are obtained as

$$r_{\text{sys}} \approx r^{\text{apprb}}_{\text{sys}} \triangleq \left\| (I - P^{\text{MTD}*}_{x_{\text{MTDc}}}) \times \Delta z^{\text{apprb}}_{\text{sys}} \right\|_2, \tag{5.31}$$

and

$$r_{\text{sys}} \approx r^{\text{apprx}}_{\text{sys}} \triangleq \left\| (I - P^{\text{MTD}*}_{x_{\text{MTDc}}}) \times \Delta z^{\text{apprx}}_{\text{sys}} \right\|_2, \tag{5.32}$$

respectively, where

$$P^{\text{MTD}*}_{x_{\text{MTDc}}} \triangleq H^{\text{MTD}*}_{x_{\text{MTDc}}} \big[(H^{\text{MTD}*}_{x_{\text{MTDc}}})^{\text{T}} H^{\text{MTD}*}_{x_{\text{MTDc}}}\big]^{-1} (H^{\text{MTD}*}_{x_{\text{MTDc}}})^{\text{T}} \tag{5.33}$$

and $H_{x_{\mathrm{MTDc}}}^{\mathrm{MTD}*} \triangleq \partial h_{\mathrm{MTD}}(x_{\mathrm{MTD}}^* + c)/\partial x$. We note that the approximation $x_{\mathrm{MTD}}^* \approx x^* + \frac{\partial x}{\partial b}|_{\Theta^*} \times \Delta b$ is adopted to guarantee the real-time property when calculating (5.31)–(5.33).

Remark 5.1 To guarantee the approximation accuracy, we adopt the sensitivities derived around two optimal points to approximate r_{att} and r_{sys}. In particular, the approximation of r_{att} is based on the sensitivities derived around the original optimal point $\Theta^* = (x^*, z, b)$, while the approximation of r_{sys} is based on the sensitivities derived around the shifted optimal point by the attacker $\Theta_{\mathrm{MTDa}}^* = \left(x_{\mathrm{MTD}}^* + c, h_{\mathrm{MTD}}(x_{\mathrm{MTD}}^* + c), b_{\mathrm{MTD}}\right)$. In practice, the injected bias vector c is not available to the system operator. Nevertheless, it will not affect the design of EXR-MTD with the derived sensitivities, as the elements in c merely act as the vulnerability factors of system states and do not have to be the true state biases injected by the attacker. Generally speaking, when designing EXR-MTD, the value of vector c is assigned in advance by the system operator. If the absolute value of the i-th element in c is significantly larger than others, then the designed EXR-MTD will make the FDI attack targeting at the i-th state more perceivable to the BDD.

(3) Numerical Analysis: In this subsection, we provide numerical results to validate the accuracy of derived approximations $r_{\mathrm{att}}^{\mathrm{appr}}$, $r_{\mathrm{sys}}^{\mathrm{apprb}}$, and $r_{\mathrm{sys}}^{\mathrm{apprx}}$. Let the set $\mathcal{L}_p$ include all branches that cover at least one PQ bus. In each scenario, we perturb one branch $(j, k) \in \mathcal{L}_p$ with the perturbation magnitude ratio $\frac{\Delta \mathbf{b}}{\mathbf{b}}$ ranging from 2% to 20% with step 2%, under which r_{att} and $r_{\mathrm{att}}^{\mathrm{appr}}$ are calculated. After introducing the FDI attack against one PQ bus covered by (j, k), where the induced voltage magnitude bias varies from 0.01p.u. to 0.1p.u., r_{sys}, $r_{\mathrm{sys}}^{\mathrm{apprb}}$, and $r_{\mathrm{sys}}^{\mathrm{apprx}}$ are computed. We first show the results in the noiseless setting and then introduce the measurement noises with standard deviations $\delta_i = 5\%, \forall 1 \leq i \leq m$ to evaluate their impact. Due to the space limitation, we merely visualize the details of residuals in IEEE 14-bus (transmission system) and radial 18-bus (distribution system) cases with $\delta_i = 0$, where 18×10 and 17×10 scenarios are depicted, respectively. According to Figs. 5.3 and 5.4, the approximation errors increase with the growth of $\Delta \mathbf{b}$ or $\mathbf{c}$, which is an intuitive phenomenon as the sensitivities are obtained around either Θ^* or Θ_{MTDa}^*. Specifically, for the residual by the attacker, $r_{\mathrm{att}}^{\mathrm{appr}}$ is very close to r_{att} especially in the radial 18-bus case. For the residual by the system operator, $r_{\mathrm{sys}}^{\mathrm{apprx}}$ has higher accuracy than $r_{\mathrm{sys}}^{\mathrm{apprb}}$ in the IEEE 14-bus case as $||\Delta \mathbf{b}||_2$ is generally larger than $||\mathbf{c}||_2$. Differently, $r_{\mathrm{sys}}^{\mathrm{apprb}}$ performs better than $r_{\mathrm{sys}}^{\mathrm{apprx}}$ in the radial 18-bus case due to the small branch ratios X/R. Finally, the essential trade-off between r_{att} and r_{sys} can be qualitatively deduced from the results. Roughly speaking, the higher r_{sys} usually leads to the comparable r_{att}, indicating that the effectiveness and hiddenness of MTD are hard to be satisfied simultaneously without a systematic design method. More details about the results under $\delta_i = 5\%$ can be found in Figs. 5.5 and 5.6.

Furthermore, the average and maximum relative approximation errors (RAEs) under $\delta_i = 0$ and $\delta_i = 5\%$ are shown in Tables 5.1 and 5.2, respectively, to validate the applicability to numerous test cases. To make the results meaningful, the metrics are counted by excluding the residuals smaller than 0.01p.u., which are indistinguishable

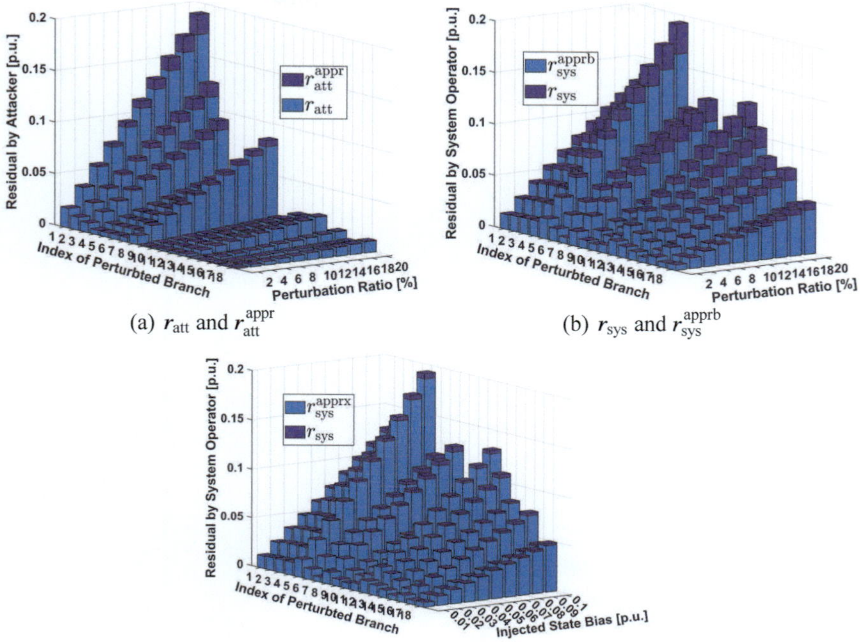

(a) r_{att} and $r_{\text{att}}^{\text{appr}}$

(b) r_{sys} and $r_{\text{sys}}^{\text{apprb}}$

(c) r_{sys} and $r_{\text{sys}}^{\text{apprx}}$

Fig. 5.3 The actual residuals r_{att} and r_{sys} and the approximated residuals $r_{\text{att}}^{\text{appr}}$, $r_{\text{sys}}^{\text{apprb}}$, and $r_{\text{sys}}^{\text{apprx}}$ when perturbing different branches in the IEEE 14-bus case (transmission system) with $\delta_i = 0$

from the impact of measurement noises. From the results, the average RAEs of $r_{\text{att}}^{\text{appr}}$ and $r_{\text{sys}}^{\text{apprx}}$ are within 6% and 4%, respectively, in transmission system test cases. Moreover, in radial 69-bus and 141-bus cases, r_{att} and $r_{\text{att}}^{\text{appr}}$ are both smaller than 0.01p.u., indicating that **the impact of reactance perturbation is neglectable**. It is noted that the RAEs of $r_{\text{sys}}^{\text{apprb}}$ and $r_{\text{sys}}^{\text{apprx}}$ show similar characteristics as those reflected by Figs. 5.3 and 5.4. **Hence, we will utilize $\mathbf{r}_{\text{att}}^{\text{appr}}$, $\mathbf{r}_{\text{sys}}^{\text{apprx}}$ for the design of EXR-MTD in power transmission systems and $\mathbf{r}_{\text{att}}^{\text{appr}}$, $\mathbf{r}_{\text{sys}}^{\text{apprb}}$ for that in power distribution systems.**

As for the impact of measurement noises, it is clearly revealed that all RAEs increase as the introduction of measurement noises. The average RAEs are still within the acceptable range (smaller than 25%), while the maximum RAEs can be extremely large. Specifically, the approximation accuracy of $r_{\text{att}}^{\text{appr}}$ is almost not affected by measurement noises in the IEEE 14-bus case as the impact of measurement noises on the residual is far smaller than that of the reactance perturbation. While in the radial 18-bus case the accuracy decreases significantly as the impact of measurement noises on the residual is comparable to or larger than that of the reactance perturbation. This phenomenon is due to that the branch ratio X/R in the IEEE 14-bus case is far larger than that in the radial 18-bus case. The approximation accuracy of $r_{\text{sys}}^{\text{apprb}}$ and $r_{\text{sys}}^{\text{apprx}}$ are not significantly affected no matter in the IEEE 14-bus

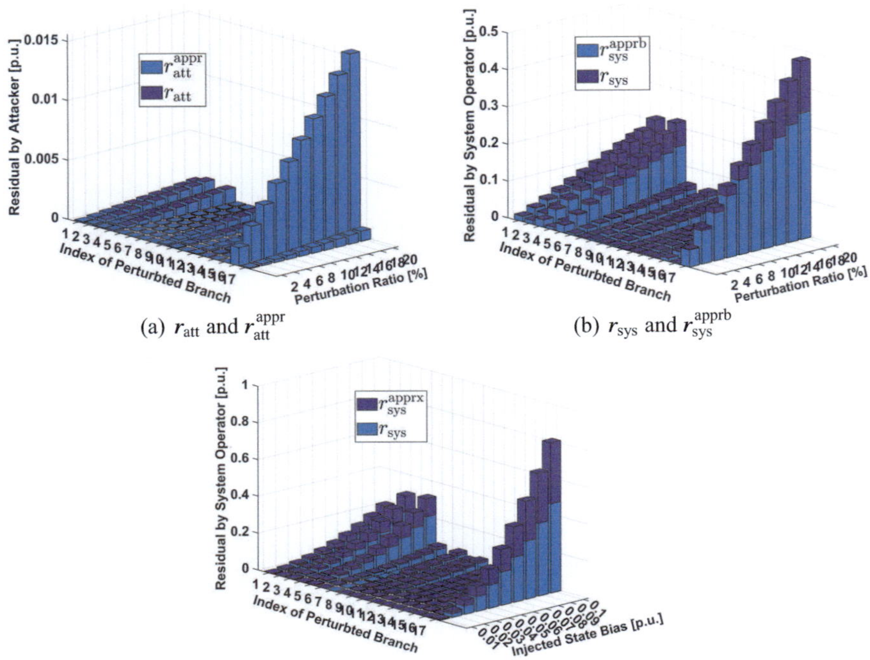

(a) r_{att} and $r_{\text{att}}^{\text{appr}}$ (b) r_{sys} and $r_{\text{sys}}^{\text{apprb}}$

(c) r_{sys} and $r_{\text{sys}}^{\text{apprx}}$

Fig. 5.4 The actual residuals r_{att} and r_{sys} and the approximated residuals $r_{\text{att}}^{\text{appr}}$, $r_{\text{sys}}^{\text{apprb}}$, and $r_{\text{sys}}^{\text{apprx}}$ when perturbing different branches in the radial 18-bus case (distribution system) with $\delta_i = 0$

case or in the radial 18-bus case as the impact of measurement noises on the residual is far smaller than that of the FDI attack and reactance perturbation. Furthermore, from the point view of effectiveness and hiddenness, the impact of measurement noises is beneficial to the hiddenness of MTD, as it can cover the impact caused by reactance perturbation. While measurement noises have a detrimental impact on the effectiveness of MTD, since it may prevent the FDI attack from being detected by BDD. **In summary, we assert that the approximations $r_{\text{att}}^{\text{appr}}$, $r_{\text{sys}}^{\text{apprb}}$, and $r_{\text{sys}}^{\text{apprx}}$ exactly capture the impact of reactance perturbation and FDI attacks on residuals. Moreover, from the above studies on the two representative cases, we can infer that measurement noises can almost guarantee the hiddenness of MTD in power distribution systems, while the hiddenness in power transmission systems and the effectiveness of MTD are almost not affected by measurement noises**. Although the impact of measurement noises on the approximated residuals has been illustrated through numerical results, it is still vital to reveal the theoretical relations between measurement noises and the approximated residuals, which is left as the future work.

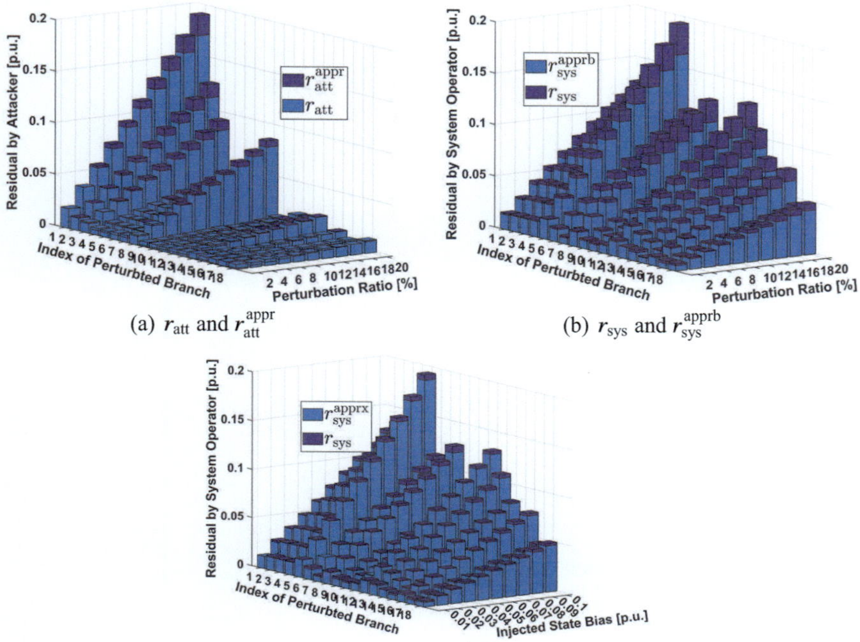

(a) r_{att} and $r_{\text{att}}^{\text{appr}}$ (b) r_{sys} and $r_{\text{sys}}^{\text{apprb}}$ (c) r_{sys} and $r_{\text{sys}}^{\text{apprx}}$

Fig. 5.5 The actual residuals r_{att} and r_{sys} and the approximated residuals $r_{\text{att}}^{\text{appr}}$, $r_{\text{sys}}^{\text{apprb}}$, and $r_{\text{sys}}^{\text{apprx}}$ when perturbing different branches in the IEEE 14-bus case (transmission system) with $\delta_i = 5\%$

5.3 Explicit Residual Based Moving Target Defense

In this section, based on the explicit approximations (5.26), (5.31)–(5.32), we propose EXR-MTD to jointly optimize the effectiveness and hiddenness.

5.3.1 Optimization Problem for Effectiveness and Hiddenness

Essentially, the design of EXR-MTD can be decomposed as the following two sub-problems:

$$\max_{\mathbf{b}_{\text{MTD}}} \ (r_{\text{sys}}^{\text{apprx}})^2 \text{ or } (r_{\text{sys}}^{\text{apprb}})^2 \qquad \text{Sub-Problem I} \tag{5.34}$$

and

$$\min_{\mathbf{b}_{\text{MTD}}} \ (r_{\text{att}}^{\text{appr}})^2, \qquad \text{Sub-Problem II} \tag{5.35}$$

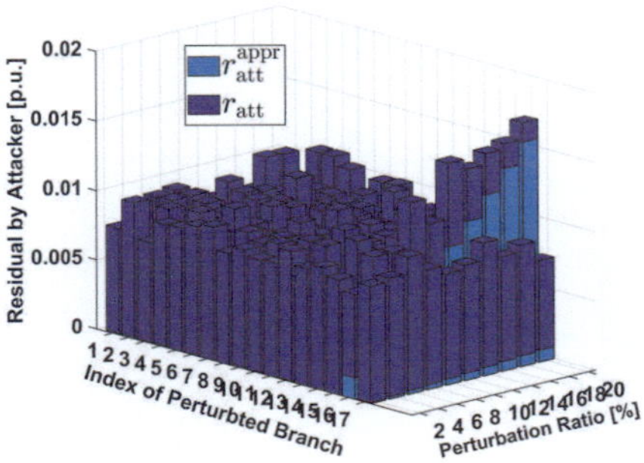

(a) r_{att} and r_{att}^{appr}

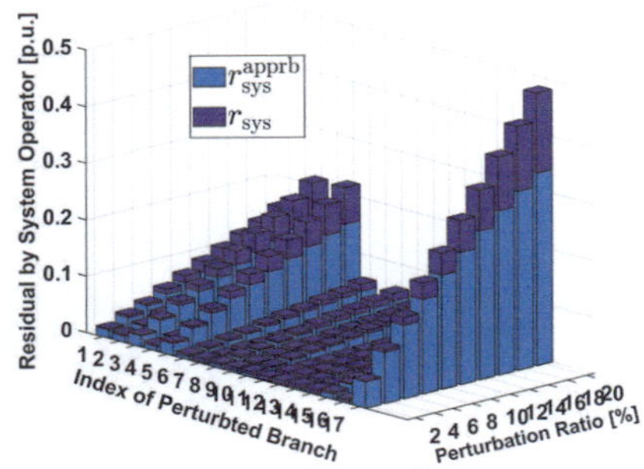

(b) r_{sys} and r_{sys}^{apprb}

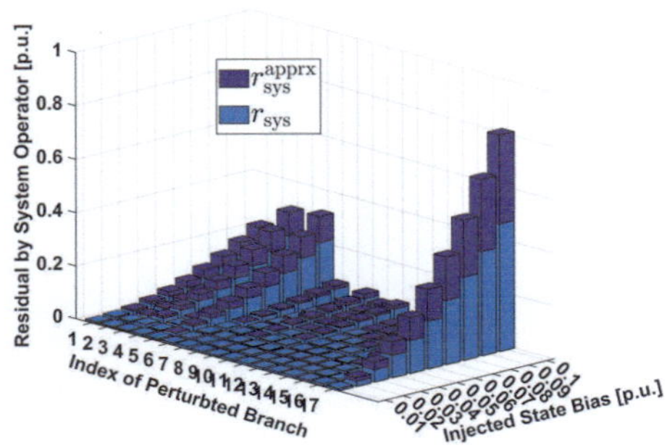

(c) r_{sys} and r_{sys}^{apprx}

Fig. 5.6 The actual residuals r_{att} and r_{sys} and the approximated residuals r_{att}^{appr}, r_{sys}^{apprb}, and r_{sys}^{apprx} when perturbing different branches in the radial 18-bus case (distribution system) with $\delta_i = 5\%$

Table 5.1 Average and maximum RAEs, $\delta_i = 0$

Test cases	Metric					
	$\lvert r_{att}^{appr} - r_{att}\rvert / r_{att}$		$\lvert r_{sys}^{apprb} - r_{sys}\rvert / r_{sys}$		$\lvert r_{sys}^{apprx} - r_{sys}\rvert / r_{sys}$	
	Avg. (%)	Max. (%)	Avg. (%)	Max. (%)	Avg. (%)	Max. (%)
IEEE 14-bus	5.58	14.54	8.85	16.74	**2.30**	4.88
IEEE 57-bus	4.77	13.02	9.52	22.70	**3.42**	27.94
IEEE 118-bus	4.84	16.81	8.99	18.14	**1.98**	9.09
IEEE 300-bus	4.49	44.77	9.77	31.24	**1.87**	31.59
Radial 18-bus	0	0	**15.75**	29.11	35.47	70.84
Radial 69-bus	–[a]	–	**7.99**	28.55	38.08	78.17
Radial 85-bus	0.72	0.76	**3.83**	12.69	25.96	80.07
Radial 141-bus	–	–	**6.75**	22.39	14.02	77.93

[a] Residuals are smaller than 0.01p.u.

Table 5.2 Average and maximum RAEs, $\delta_i = 5\%$

Test cases	Metrics					
	$\lvert r_{att}^{appr} - r_{att}\rvert / r_{att}$		$\lvert r_{sys}^{apprb} - r_{sys}\rvert / r_{sys}$		$\lvert r_{sys}^{apprx} - r_{sys}\rvert / r_{sys}$	
	Avg. (%)	Max. (%)	Avg. (%)	Max. (%)	Avg. (%)	Max. (%)
IEEE 14-bus	5.53	19.21	10.12	22.13	**3.88**	19.84
IEEE 57-bus	22.82	97.31	18.31	71.80	**12.93**	81.39
IEEE 118-bus	25.93	99.75	12.25	93.83	**5.69**	99.32
IEEE 300-bus	29.96	99.99	29.37	99.96	**23.56**	99.99
Radial 18-bus	16.12	25.54	**20.18**	31.73	27.94	72.23
Radial 69-bus	–	–	**12.28**	30.37	37.68	88.17
Radial 85-bus	48.28	52.07	**10.25**	54.26	31.15	85.04
Radial 141-bus	–	–	**12.59**	62.26	18.29	87.93

where[2] (5.34) and (5.35) represent the maximization of effectiveness and hiddenness, respectively. Here the main challenge lies in the matrix inverse issue involved in r_{sys}^{apprx} and r_{sys}^{apprb}, which can significantly enlarge the computation complexity of sub-problem I and even make the (local) optimum unavailable. Hence, we first make some transformations for sub-problem I to address the matrix inverse issue, after which the two sub-problems are aggregated together by multiplying an appropriate weight.

5.3.2 *Problem Transformation and Theoretical Analysis*

To address the matrix inverse issue in sub-problem I, we deeply investigate the physical meaning behind matrix $I - P_{\mathbf{x}_{MTDc}}^{MTD*}$, which is named as the *projection matrix* in the field of linear regression [34] as it orthogonally projects vectors onto the subspace $\mathcal{R}_{\mathbf{x}}^{\perp}$ that is orthogonal to the image of $H_{\mathbf{x}_{MTDc}}^{MTD*}$. For convenience, the approximation of $\Delta\mathbf{z}_{sys}$ is omitted, under which (5.34) can be uniformly described by

$$\max_{\mathbf{b}_{MTD}} \underbrace{\left\| (I - P_{\mathbf{x}_{MTDc}}^{MTD*}) \times \Delta\mathbf{z}_{sys} \right\|_2^2}_{a^2} \tag{5.36}$$

$$\Updownarrow$$

$$\max_{\mathbf{b}_{MTD}} \underbrace{\left\| \Delta\mathbf{z}_{sys} \right\|_2^2}_{c^2} - \underbrace{\left\| P_{\mathbf{x}_{MTDc}}^{MTD*} \times \Delta\mathbf{z}_{sys} \right\|_2^2}_{b^2}, \tag{5.37}$$

[2] $(r_{sys}^{apprx})^2$ is utilized for power transmission systems and $(r_{sys}^{apprb})^2$ is for power distribution systems.

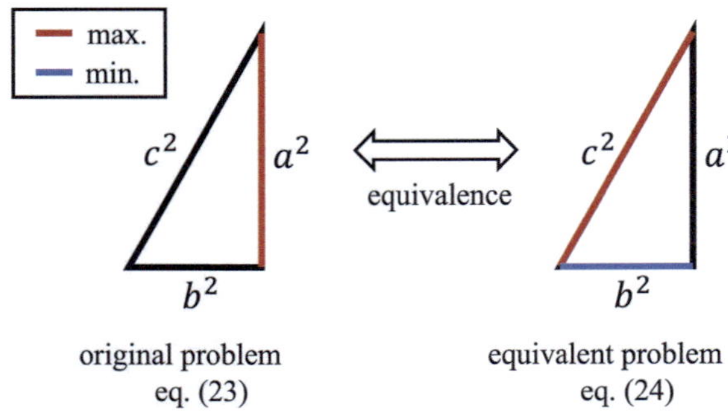

Fig. 5.7 Graphical illustration of the equivalence (5.36) ⇔ (5.37)

where (5.37) is an equivalent formation of (5.36) and the equivalence can be graphically interpreted via a right triangle. Let $(H_{\mathbf{x}_{\mathrm{MTDc}}}^{\mathrm{MTD}*})^{\perp} \in \mathbb{R}^{m\times(m-n)}$ be a full column rank matrix and satisfy $(H_{\mathbf{x}_{\mathrm{MTDc}}}^{\mathrm{MTD}*})^{\mathrm{T}}(H_{\mathbf{x}_{\mathrm{MTDc}}}^{\mathrm{MTD}*})^{\perp} = 0$. Then, the column vectors of matrices $H_{\mathbf{x}_{\mathrm{MTDc}}}^{\mathrm{MTD}*}$ and $(H_{\mathbf{x}_{\mathrm{MTDc}}}^{\mathrm{MTD}*})^{\perp}$ constitute a basis for the m-dimensional vector space. For any vector $\Delta\mathbf{z}_{\mathrm{sys}} \in \mathbb{R}^m$, there exist $\mathbf{h}_1 \in \mathbb{R}^n$ and $\mathbf{h}_2 \in \mathbb{R}^{m-n}$ such that $\Delta\mathbf{z}_{\mathrm{sys}} = H_{\mathbf{x}_{\mathrm{MTDc}}}^{\mathrm{MTD}*} \times \mathbf{h}_1 + (H_{\mathbf{x}_{\mathrm{MTDc}}}^{\mathrm{MTD}*})^{\perp} \times \mathbf{h}_2$, whose orthogonal projection onto image of $H_{\mathbf{x}_{\mathrm{MTDc}}}^{\mathrm{MTD}*}$ is captured with

$$P_{\mathbf{x}_{\mathrm{MTDc}}}^{\mathrm{MTD}*} \times \Delta\mathbf{z}_{\mathrm{sys}} = H_{\mathbf{x}_{\mathrm{MTDc}}}^{\mathrm{MTD}*} \times \mathbf{h}_1. \tag{5.38}$$

Hence, the three variables a, b, and c can be regarded as the three sides of a right triangle, where a and b are the two sides adjacent to the right angle and c denotes the hypotenuse. The equivalence of (5.36) ⇔ (5.37) is vividly demonstrated in Fig. 5.7.

Proposition 5.2 is established to avert the matrix inverse issue involved in (5.37), where maximizing $-b^2$ is transformed to maximizing the lower bound.

Proposition 5.2 *Based on matrix norm inequalities, maximizing* (5.37) *is transformed to maximizing the lower bound, i.e.,*

$$\max_{\boldsymbol{b}_{MTD}} \quad \left\| \Delta z_{sys} \right\|_2^2 - \frac{\left\| (H_{\boldsymbol{x}_{MTDc}}^{MTD*})^{\mathrm{T}} \times \Delta z_{sys} \right\|_2^2}{\left(\sigma_{min}(H_{\boldsymbol{x}_{MTDc}}^{MTD*})\right)^2}, \tag{5.39}$$

where $\sigma_{min}(\cdot)$ denotes the minimal singular value function.

Proof ollowing (5.37) and (5.38), the length of the side signified by b satisfies

$$b^2 = \left\| H_{\boldsymbol{x}_{\mathrm{MTDc}}}^{\mathrm{MTD}*} \times \boldsymbol{h}_1 \right\|_2^2.$$

Moreover, with $\Delta z_{sys} = H_{\boldsymbol{x}_{MTDc}}^{MTD*} \times \boldsymbol{h}_1 + (H_{\boldsymbol{x}_{MTDc}}^{MTD*})^{\perp} \times \boldsymbol{h}_2$, we have

$$\begin{aligned}
&\left\|(H_{\boldsymbol{x}_{MTDc}}^{MTD*})^T \times \Delta \boldsymbol{z}_{sys}\right\|_2^2 = \left\|(H_{\boldsymbol{x}_{MTDc}}^{MTD*})^T \times H_{\boldsymbol{x}_{MTDc}}^{MTD*} \times \boldsymbol{h}_1\right\|_2^2 \\
&= (H_{\boldsymbol{x}_{MTDc}}^{MTD*} \times \boldsymbol{h}_1)^T \times H_{\boldsymbol{x}_{MTDc}}^{MTD*} \times (H_{\boldsymbol{x}_{MTDc}}^{MTD*})^T \times H_{\boldsymbol{x}_{MTDc}}^{MTD*} \times \boldsymbol{h}_1 \\
&= \sum_{i=1}^{n} \lambda_i \times \alpha_i^2 \times \|\boldsymbol{v}_i\|_2^2,
\end{aligned} \tag{5.40}$$

where λ_i and $\boldsymbol{v}_i, \forall i \in \{1, \ldots, n\}$ represent positive eigenvalues and corresponding unit orthogonal eigenvectors of matrix $(H_{\boldsymbol{x}_{MTDc}}^{MTD*})^T \times H_{\boldsymbol{x}_{MTDc}}^{MTD*}$, respectively. Here eigenvalues are ordered in a non-decreasing manner, and parameters α_i satisfy

$$H_{\boldsymbol{x}_{MTDc}}^{MTD*} \times \boldsymbol{h}_1 = \sum_{i=1}^{n} \alpha_i \times \boldsymbol{v}_i.$$

Hence, by replacing all λ_i in (5.40) with λ_1, we obtain

$$\left\|(H_{\boldsymbol{x}_{MTDc}}^{MTD*})^T \times \Delta \boldsymbol{z}_{sys}\right\|_2^2 \geq \lambda_1 \times \sum_{i=1}^{n} \alpha_i^2 = \lambda_1 \times b^2, \tag{5.41}$$

which is a upper bound of b^2. The result follows after substituting (5.41) into (5.37) as $\sigma_{\min}(H_{\boldsymbol{x}_{MTDc}}^{MTD*}) = \sqrt{\lambda_1}$.

Remark 5.2 Due to the existence of $\sigma_{\min}(\cdot)$, it is still challenging to directly solve (5.39). Fortunately, through numerical simulations, we observe that the impact of reactance perturbation on $\left(\sigma_{\min}(H_{\mathbf{x}_{MTDc}}^{MTD*})\right)^2$ is neglectable, which may be further utilized to simplify (5.39). We simulate all scenarios where the reactance of one branch is perturbed with ratio 20%, and the average and maximum relative alterations of $\left(\sigma_{\min}(H_{\mathbf{x}_{MTDc}}^{MTD*})\right)^2$ under 8 test cases are presented in Table 5.3. The average relative alterations are all 0, and the maximum relative alterations are within 3%. Hence, we consider that it is feasible to treat $\left(\sigma_{\min}(H_{\mathbf{x}_{MTDc}}^{MTD*})\right)^2$ as constant in the presence of MTD when the perturbation magnitude ratio is bounded by 20% and (5.39) is simplified as

$$\max_{\mathbf{b}_{MTD}} \quad \left\|\Delta \mathbf{z}_{sys}\right\|_2^2 - \frac{\left\|(H_{\mathbf{x}_{MTDc}}^{MTD*})^T \times \Delta \mathbf{z}_{sys}\right\|_2^2}{\left(\sigma_{\min}(H_{\mathbf{x}_c}^*)\right)^2}, \tag{5.42}$$

where $H_{\mathbf{x}_c}^* \triangleq \partial \mathbf{h}(\mathbf{x}^* + \mathbf{c})/\partial \mathbf{x}$.

Table 5.3 Average and maximum relative alterations on $\sigma_{\min}^{a}$

Test cases	Metrics		Test cases	Metrics	
	$\Delta\sigma_{\min}^{b}/\sigma_{\min,0}^{c}$			$\Delta\sigma_{\min}/\sigma_{\min,0}$	
	Avg.	Max. (%)		Avg.	Max. (%)
IEEE 14-bus	0[d]	0.09	Radial 18-bus	0	0.64
IEEE 57-bus	0	0	Radial 69-bus	0	0.09
IEEE 118-bus	0	0.09	Radial 85-bus	0	0.08
IEEE 300-bus	0	2.33	Radial 141-bus	0	0.12

[a] $\left(\sigma_{\min}(H_{\mathbf{x}_{\mathrm{MTDc}}}^{\mathrm{MTD}*})\right)^2$ is written as $\sigma_{\min}$ for brevity.
[b] $\Delta\sigma_{\min}$ denotes the absolute alteration on $\sigma_{\min}$ caused by MTD.
[c] $\sigma_{\min,0}$ denotes the $\sigma_{\min}$ in the absence of MTD.
[d] The value is considered to be 0 if it is smaller than 0.0001.

(1) Aggregation of Sub-problems: By aggregating sub-problems (5.35) and (5.42), we obtain

$$\begin{aligned}\min_{\mathbf{b}_{\mathrm{MTD}}} \; & -\left\|\Delta\mathbf{z}_{\mathrm{sys}}\right\|_2^2 + \frac{\left\|(H_{\mathbf{x}_{\mathrm{MTDc}}}^{\mathrm{MTD}*})^{\mathrm{T}} \times \Delta\mathbf{z}_{\mathrm{sys}}\right\|_2^2}{\left(\sigma_{\min}(H_{\mathbf{x}_{\mathrm{c}}}^{*})\right)^2} + w_{\mathrm{att}} \times \left\|\Delta\mathbf{z}_{\mathrm{att}}^{\mathrm{appr}}\right\|_2^2 \\ \text{s.t.} \quad & \mathbf{b} + \Delta\underline{\mathbf{b}} \leq \mathbf{b}_{\mathrm{MTD}} \leq \mathbf{b} + \Delta\overline{\mathbf{b}},\end{aligned} \tag{5.43}$$

where $w_{\mathrm{att}} \geq 1$ is the weight parameter that should be appropriately chosen to maximize the effectiveness of MTD while guaranteeing the hiddenness [18]. Specifically, a larger w_{att} leads to the designed EXR-MTD possessing stronger hiddenness, and a smaller w_{att} results in the opposite result.

After substituting (5.29) and (5.30) into (5.43), the nonlinear and non-convex optimization problems can be obtained for power distribution and power transmission systems, respectively, which can be treated as the polynomial optimization problems (POPs) and are NP-hard. It is difficult to find the global optimums of the formulated PoPs, and further efforts are still required to develop approximation algorithms to approach them [35], which is left as the future work. In this study, we use the standard *fmincon* solver from Matlab equipped with the interior-point algorithm to solve the PoPs, where different local minimums could be found when started from different initial points. Since the larger perturbation magnitude usually makes the MTD possess stronger effectiveness, we consider the two boundary points, i.e., $\mathbf{b} + \Delta\underline{\mathbf{b}}$ and $\mathbf{b} + \Delta\overline{\mathbf{b}}$, as initial points, under which two (local) minimums can be obtained and the one with better performance is chosen as the output.

Finally, we note that the EXR-MTD is dedicated to enhancing the cybersecurity of SE against FDI attacks based on the D-FACTS devices that have been installed to branches for the power flow regulation. While the installation of extra D-FACTS devices is not necessary for the implementation of EXR-MTD.

(2) Numerical Results: In this section, we conduct extensive simulations on the test cases extracted from MATPOWER to evaluate the performance of EXR-MTD.

The standard deviations of real-time measurement noises are considered to be 1% for phasor and magnitude measurements. The pseudo measurement noises are considered to possess 10% standard deviations. The perturbation magnitude ratios are bounded by 20%, under which the power loss variations would be restricted within 1% in a sufficiently large number of perturbation cases [9]. When solving problem (5.43), the weight parameter w_{att} is set as 100 for power transmission systems and 1 for power distribution systems, and moreover all elements of vector $\mathbf{c}$ are set as 0.1. The significance level α utilized in BDD is set as 0.05. For the purpose of practicality, we assume that both the attacker and the system operator adopt the AC power flow model.

5.4 Performance Evaluation

In this section, we conduct extensive simulations on the test cases extracted from MATPOWER to evaluate the performance of EXR-MTD. The standard deviations of real-time measurement noises are considered to be 1% for phasor and magnitude measurements. The pseudo measurement noises are considered to possess 10% standard deviations. The perturbation magnitude ratios are bounded by 20%, under which the power loss variations would be restricted within 1% in a sufficiently large number of perturbation cases [9]. When solving problem (5.43), the weight parameter w_{att} is set as 100 for power transmission systems and 1 for power distribution systems, and moreover all elements of vector $\boldsymbol{c}$ are set as 0.1. The significance level α utilized in BDD is set as 0.05. For the purpose of practicality, we assume that both the attacker and the system operator adopt the AC power flow model.

5.4.1 Explicit Residual Based Moving Target Defense in Transmission Systems

In this subsection, we compare the performance of EXR-MTD with that of PFI-MTD [14], DC-HMTD [18], and AC-HMTD [18]. In particular, the PFI-MTD is to make the power flow invariant in the simplified DC model, and the DC-HMTD and AC-HMTD are to optimize the effectiveness and hiddenness of MTD based on the DC and AC models, respectively.

EXR-MTD and PFI-MTD

The PFI-MTD is implemented with the perturbation magnitude ratio of at least one branch reaching 0.2 or −0.2 to maximize the effectiveness. First, when all branches can be perturbed, we depict residuals under EXR-MTD and PFI-MTD,

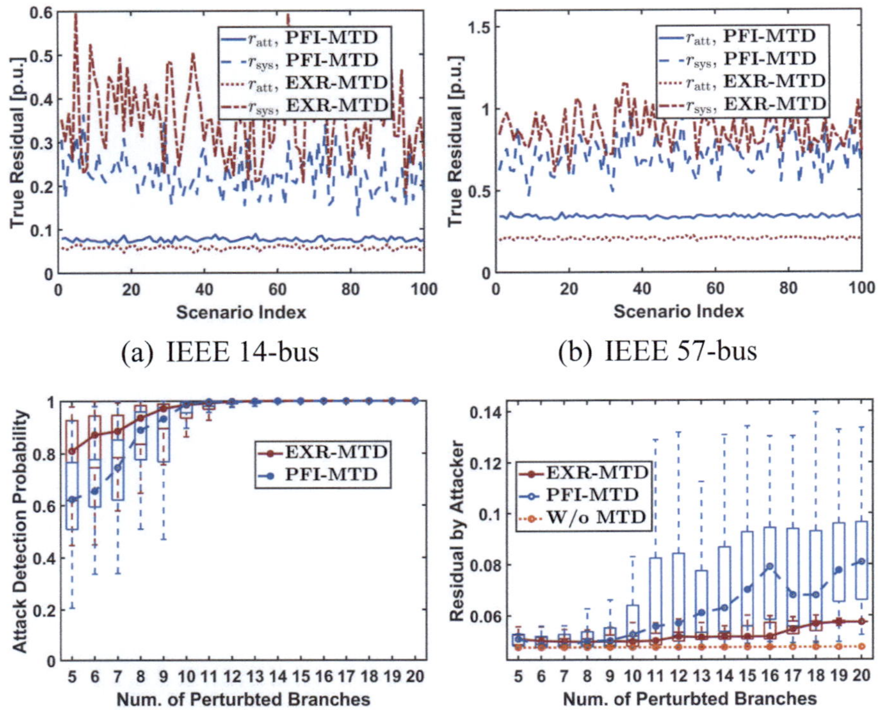

(a) IEEE 14-bus (b) IEEE 57-bus

(c) IEEE 14-bus

Fig. 5.8 The comparison results between EXR-MTD and PFI-MTD [14] in transmission systems, where (**a**) and (**b**) depict the true residuals when all branches can be perturbed, and (**c**) shows the effectiveness and hiddenness of MTD when the number of perturbed branches increases

and consider 100 scenarios. In each scenario, we sample each element of $\boldsymbol{c}$ from a uniform distribution $\mathcal{U}(-dm, dm)$ with $dm = 0.1$ being the maximum magnitude of the injected biases into state variables. As shown in (a) and (b) of Fig. 5.8, compared with PFI-MTD, EXR-MTD can lead to the smaller r_{att}, and moreover under more than 90% scenarios, the larger r_{sys} is induced. Then, we consider the case where only partial branches can be perturbed. According to [14], the number of perturbed branches ranges from 5 to 20 in the IEEE 14-bus case. For each setting, 1000 attacks generated from $\mathcal{U}(-0.1, 0.1)$ are launched, and based on Monte Carlo simulations, the attack detection probability is estimated as $\frac{\text{Num. of detected FDIAs}}{1000}$. Besides, 100 PFI-MTD strategies are randomly selected for each number of perturbed branches, and 100 EXR-MTD strategies are correspondingly designed with the same sets of perturbed branches. As illustrated in (c) of Fig. 5.8, the attack detection probability becomes higher as the number of perturbed branches increases, and EXR-MTD generally performs better than PFI-MTD in detecting FDI attacks. Additionally, it is observed that PFI-MTD is difficult to maintain hiddenness when the attacker uses the AC power flow model, as the induced r_{att} is significantly larger than that without MTD. On the contrary, EXR-MTD has the evident advantage in preserving hiddenness since the incurred r_{att} is close to that without MTD. From the above results, it is asserted that EXR-MTD comprehensively performs better than PFI-MTD when the practical AC power flow model is adopted, despite the utilization of approximated residuals.

EXR-MTD, DC-HMTD, and AC-HMTD

Based on the HMTD planning solutions in IEEE 14-bus and 57-bus cases [18], we solve the optimization problems formulated for EXR-MTD, DC-HMTD, and AC-HMTD and compare the effectiveness and hiddenness of obtained MTD strategies. In particular, the optimization problems are all solved by the standard *fmincon* function from Matlab equipped with the interior-point algorithm, in which the bound of perturbation magnitude ratio varies from 2% to 20%. Moreover, when solving the optimization problems, the initial points are chosen from the two boundary points, i.e., $\boldsymbol{b} + \Delta\underline{\boldsymbol{b}}$ and $\boldsymbol{b} + \Delta\overline{\boldsymbol{b}}$. The results indicate that the effectiveness increases with the growth of the perturbation bound, while the hiddenness continuously decreases. Specifically, the effectiveness of EXR-MTD is slightly weaker than those of DC-HMTD and AC-HMTD when the perturbation bound is smaller than 12.5%, and the attack detection probabilities can all reach 1 if the perturbation bound exceeds 12.5%. Moreover, the hiddenness of EXR-MTD is always stronger than those of DC-HMTD and AC-HMTD. The computation time of solving the three optimization problems in a core i9 computer, which has a 3.6GHz CPU and 32.0G memory, is listed in Table 5.4. In each test case, the problem is solved 100 times, and the average and maximum computation time is demonstrated. It is shown that the computation time of EXR-MTD is longer than that of DC-HMTD and is shorter than that of AC-HMTD.

Remark 5.3 Compared with DC-HMTD, the design of EXR-MTD considers the existence of series resistances and charging capacitances at power lines, and adopts sensitivity analysis to linearly approximate the ACPF model and obtain the approximated measurement residuals (Fig. 5.9). Indeed, the approximation error increases with the growth of the perturbation magnitude. Nevertheless, in IEEE 14-bus and

Table 5.4 Computation time [s]

Test cases	Metric					
	DC-HMTD		AC-HMTD		EXR-HMTD	
	Avg.	Max.	Avg.	Max.	Avg.	Max.
IEEE 14-bus	0.0260	0.0322	0.8723	1.0983	0.4233	0.5280
IEEE 57-bus	0.2691	0.6282	9.9994	10.205	6.9812	8.0752

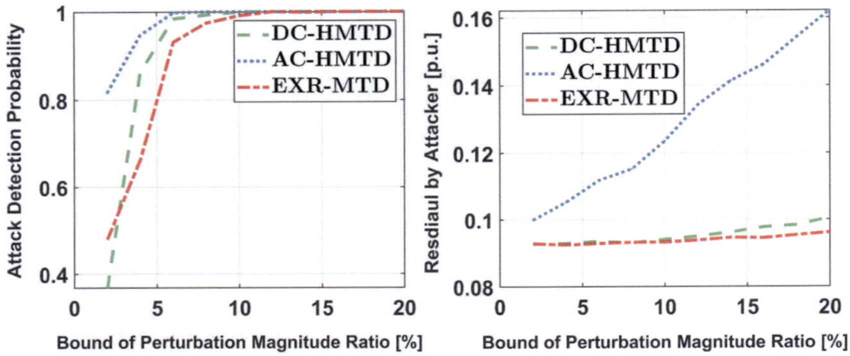

(a) IEEE 14-bus

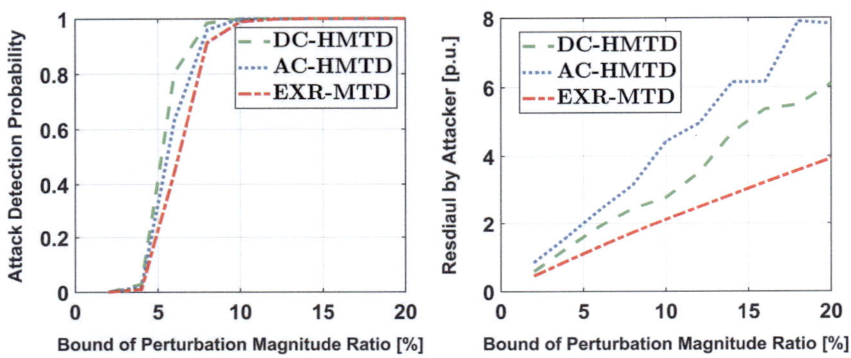

(b) IEEE 57-bus

Fig. 5.9 The effectiveness and hiddenness of EXR-MTD in comparison with those of DC-HMTD and AC-HMTD, where (**a**) and (**b**) show the results in IEEE 14-bus and 57-bus cases, respectively

57-bus cases, the average branch ratios R/X are 39.83% and 33.90%, respectively, which are both larger than the maximal perturbation magnitude ratio on branch reactances, i.e., 20%. Intuitively, although the accuracy of the approximated PF model decreases as the increase of the perturbation magnitude, the induced distortions are always smaller than those caused by the neglected branch resistances as in the DCPF model. Hence, the hiddenness of EXR-MTD can be stronger than that of DC-HMTD, and the stronger hiddenness will restrain the perturbation magnitude, under which the effectiveness of EXR-MTD can be slightly weaker than that of DC-HMTD when the perturbation bound is smaller than 12.5%. Once the perturbation bound exceeds 12.5%, the induced residuals will both trigger the attack alarm, and the attack detection probabilities are equal to 1.

Remark 5.4 Moreover, since the AC-HMTD directly minimizes the ACPF variations caused by reactance perturbation, more decision variables such as the voltage magnitudes and angles after reactance perturbation are included in the formulated optimization problem. Given a non-convex optimization problem and the solver, the selection of the initial values of decision variables determines how good the attained local minimum is. The increase of the number of decision variables would make the selection of a good initial point more difficult and time-consuming. Thus, when using the voltage magnitudes and angles without reactance perturbation as initial values and adopting the standard solver *fmincon* from Matlab equipped with the interior-point algorithm, the hiddenness of obtained AC-HMTD can be weaker than that of EXR-MTD when the effectiveness of AC-HMTD is comparable to that of EXR-MTD. Furthermore, the increase of the number of decision variables would also enlarge the matrix dimensions involved in the interior-point algorithm when solving the optimization problem, under which the computation time of AC-HMTD can be longer than those of DC-HMTD and EXR-MTD. Moreover, the computation time of DC-HMTD is shorter than that of EXR-MTD as the order of the formulated polynomial optimization problem (PoP) for DC-HMTD (i.e., 2) is smaller than that of EXR-MTD (i.e., 6).

5.4.2 *Explicit Residual Based Moving Target Defense in Distribution Systems*

In this subsection, we investigate the performance of EXR-MTD in power distribution systems. First, we show the residuals r_{att} and r_{sys} in radial 18-bus and 69-bus cases, when all branches can be perturbed. The results depicted in (a) and (b) of Fig. 5.10 indicate that it is hard for the attacker to identify the existence of EXR-MTD as r_{att} is almost invariant after implementing it. Meanwhile, EXR-MTD can effectively expose the FDI attacks constructed with outdated branch parameters to the BDD due to the noticeable improvement of r_{sys}. Then, extensive simulations are conducted in the radial 18-bus case to validate the performance of EXR-MTD when merely partial branches can be perturbed. For each number of perturbed branches, 100 branch sets are randomly selected. Clearly, the attack detection probability increases when more branches are perturbed, i.e., more buses are covered by perturbed branches. In particular, we find that the attack detection probability is around 99% when more than 10 branches are perturbed. If the number of perturbed branches is 5, the attack detection probability can be smaller than 60% as merely 6 buses are covered. It is also worthwhile noting that r_{att} increases as the growth of the number of perturbed branches, but fortunately the discrimination compared with that without EXR-MTD is among 0.001p.u. Furthermore, it is observed that the larger standard deviations of measurement noises will decrease the attack detection probability as a larger detection threshold is required to tolerate the fluctuations caused by measurement noises. Oppositely, the hiddenness of EXR-MTD is improved because more impact of reactance perturbation on r_{att} can be covered by that of measurement noises.

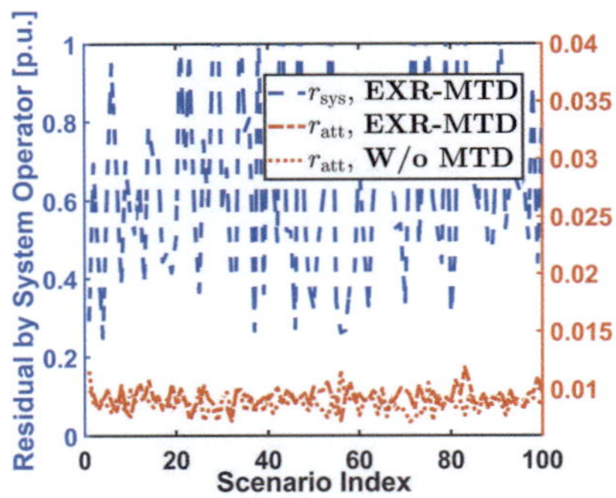

(a) Radial 18-bus

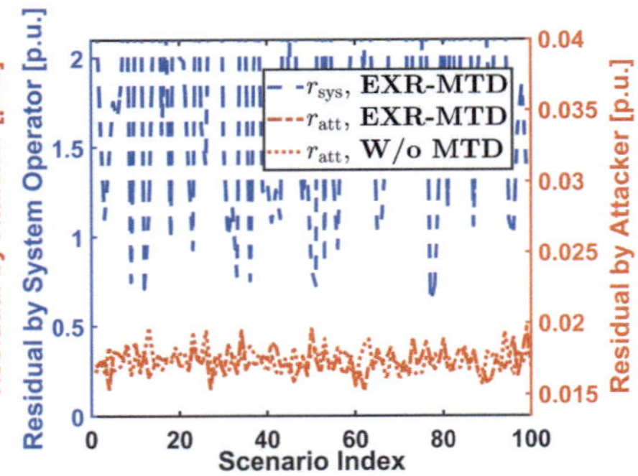

(b) Radial 69-bus

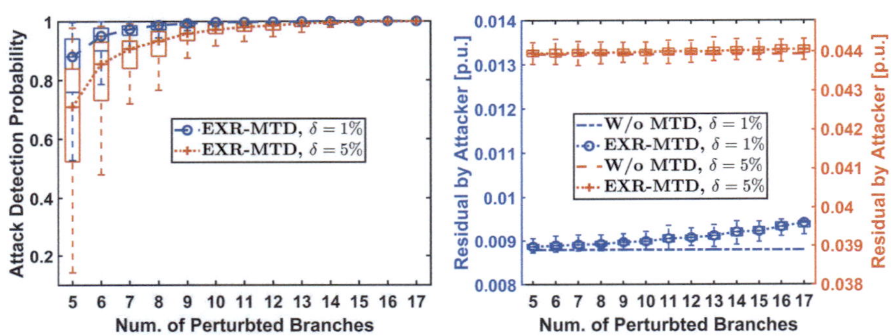

(c) Radial 18-bus

Fig. 5.10 The results of EXR-MTD in power distribution systems, where (**a**) and (**b**) depict the true residuals when all branches can be perturbed, and (**c**) shows the effectiveness and hiddenness of EXR-MTD when the number of perturbed branches and δ_i increase

5.4.3 Trade-Off Between Effectiveness and Hiddenness

In this subsection, we show the impact of weight parameter w_{att} on the effectiveness and hiddenness of EXR-MTD, where 10 logarithmically equally spaced points are selected between 10^0 and 10^4 as illustrated in Fig. 5.11. Evidently, in the IEEE 14-bus case, r_{att} and r_{sys} both decrease as the growth of w_{att}, indicating that w_{att} plays a vital role in balancing the trade-off between the two metrics. Specifically, the larger w_{att} means the stronger hiddenness and the weaker effectiveness of EXR-MTD. Moreover, the same phenomenon appears in the radial 18-bus case except that the variation of r_{att} is within 0.001p.u. Hence, we choose $w_{\text{att}} = 100$ for power transmission systems to guarantee that r_{att} is not too large, and $w_{\text{att}} = 1$ for power distribution systems to maximize r_{sys}.

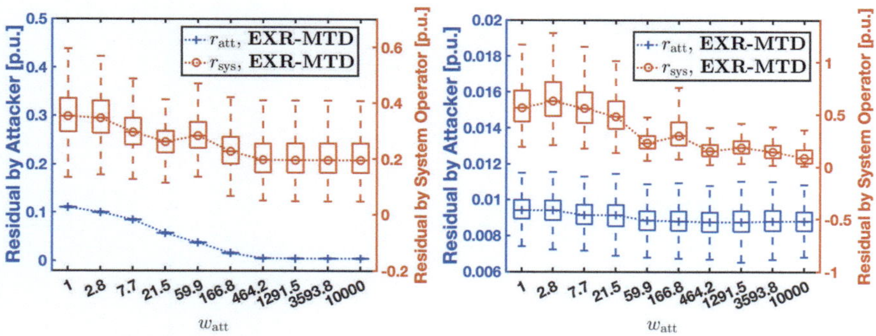

(a) IEEE 14-bus (b) Radial 18-bus

Fig. 5.11 The impact of weight w_{att} on the effectiveness and hiddenness of EXR-MTD

5.4.4 Analysis of Generation Cost Impact, Computation Time, and Scalability

Generation Cost Impact: In this subsection, we show the impact of EXR-MTD on the OPF generation cost, where weight parameter w_{att} and the bound of perturbation magnitude ratio vary as shown in Fig. 5.12. It is noticed that the alteration of OPF generation cost is within 1.5\$/hr in the IEEE 14-bus case, and bounded by 0.01\$/hr in the radial 18-bus case. Though the OPF generation cost is not directly considered in (5.43), we observe that the alteration of OPF generation cost is reduced with the growth of w_{att}, under which the designed EXR-MTD has the stronger hiddenness and the fluctuations of AC power flows caused by MTD is more limited. Hence, to

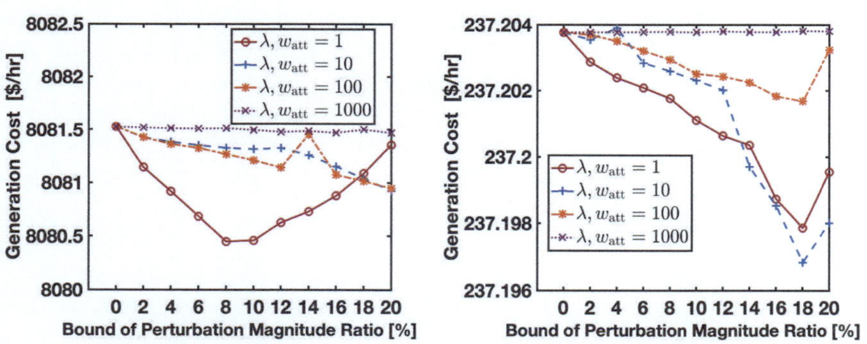

(a) IEEE 14-bus (b) Radial 18-bus

Fig. 5.12 The OPF generation cost under EXR-MTD when weight parameter w_{att} and the bound of perturbation magnitude ratio vary

Table 5.5 Computation time [s]

Test cases	Metrics		Test cases	Metrics	
	Avg.	Max.		Avg.	Max.
IEEE 14-bus	0.6028	1.0018	Radial 18-bus	0.7928	1.4860
IEEE 57-bus	11.075	13.934	Radial 69-bus	0.6836	1.0854
IEEE 118-bus	88.900	91.533	Radial 85-bus	0.8289	1.3299
IEEE 300-bus	674.32	762.02	Radial 141-bus	4.0657	5.1992

reduce the impact of EXR-MTD on the OPF generation cost, one alternative method is to increase w_{att} when designing EXR-MTD.

Computation Time: In this subsection, we evaluate the computation time of solving (5.43) in the 8 test cases extracted from Matpower when all branches can be perturbed and show the results in Table 5.5. Similarly, the problem is solved 100 times in the core i9 computer, and the average and maximum computation time is illustrated. We find that the computation time of designing EXR-MTD in power distribution systems is in the order of seconds and can be neglectable. While the computation time of designing EXR-MTD in power transmission systems increases with the number of buses and can reach at most 12.7 minutes in the IEEE 300-bus test case, which is feasible as the perturbation commands are updated hourly in practice [16].

Scalability to Large-scale Systems: In this subsection, we test the scalability of EXR-MTD via 4 polish system test cases (including more than 2000 buses) extracted from MATPOWER. When designing EXR-MTD, we assume that all branches are equipped with D-FACTS devices and can be perturbed. The computation time and the performance of obtained EXR-MTD (reflected by r_{att} and r_{sys}) are illustrated in Table 5.6. It is observed that when the weight parameter ω_{att} is appropriately chosen, then the EXR-MTD strategy with strong effectiveness and hiddenness can be quickly obtained, where r_{sys} is significantly larger than r_{att}. Nevertheless, if ω_{att} is not well tuned, then the computation time of solving (5.43) can exceed 6 hours, which is

Table 5.6 Results in polish system test cases

Test cases	Metric			
	ω_{att}	r_{att} [p.u.]	r_{sys} [p.u.]	Time [s]
Case2383wp	50	2.7520	4.0290	1289.6
Case2737sop	200	1.2409	7.1925	725.38
Case3012wp	1	2.1738	22.892	864.92
Case3120sp	1	2.7163	22.022	980.19

caused by the operations of large-scale sparse matrices. Hence, in the future work, an efficient and dedicated solver is required to address this issue.

5.4.5 Discussions

In this subsection, we discuss the differences when applying EXR-MTD to power transmission and power distribution systems. We find that the EXR-MTD in power distribution systems performs better than that in power transmission systems in terms of hiddenness and effectiveness, which are illustrated individually as follows.

Effectiveness: The effectiveness of MTD in detecting FDI attacks is directly related to the number of covered buses by perturbed branches [9]. Simply, if there exist buses not covered by perturbed branches, then the FDI attacks against these buses can still easily bypass BDD. In Fig. 5.13, we compare the bus coverage ratios in IEEE 14-bus and radial 18-bus cases. For each number of perturbed branches, EXR-MTD is designed under 100 randomly selected branch sets, and the number of the buses covered by the branches whose perturbation magnitude ratios are larger than 0.1% is counted. The bus coverage ratio is computed via dividing the number of covered buses by the number of total buses excluding the reference bus. It is observed that the bus coverage ratio in the radial 18-bus case is generally larger than that in the IEEE 14-bus case. Hence, when the numbers of the branches equipped with D-FACTS devices are the same, the EXR-MTD in the radial 18-bus case is likely to induce higher attack detection probability as demonstrated by Figs. 5.8 and 5.10. Two reasons are presented to explain the phenomenon. First, the impact of reactance perturbation on r_{att} in power distribution systems is small and even can be neglectable as indicated by Table 5.1, and thus more branches can be perturbed therein compared with those in power transmission systems when guaranteeing the same value of r_{att}.

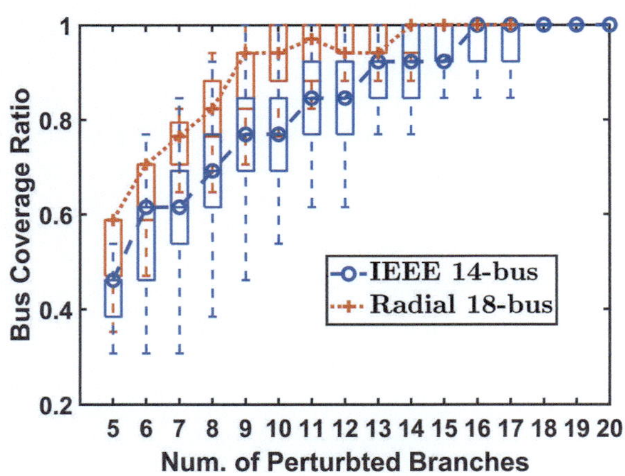

Fig. 5.13 The bus coverage ratios of EXR-MTD in IEEE 14-bus and radial 18-bus cases

Secondly, the *radial* network is also helpful to improve the bus coverage ratio as the same number of branches therein is likely to cover more buses compared with that in the *mesh* network.

Hiddenness: According to Figs. 5.3 and 5.4, the impacts of reactance perturbation on measurement residual r_{att} in power transmission and power distribution systems are totally different. Specifically, perturbing any single branch in the IEEE 14-bus case (transmission system) is likely to induce $r_{\text{att}} > 0.02$p.u., while $r_{\text{att}} < 0.016$p.u. can be always guaranteed in the radial 18-bus case (distribution system). Moreover, the hiddenness of EXR-MTD behaves similarly in IEEE 14-bus and radial 18-bus cases as indicated by (c) of Fig. 5.8 and (c) of Fig. 5.10, respectively. Concretely, the EXR-MTD in the IEEE 14-bus case can improve r_{att} by at least 0.01p.u., while in the radial 18-bus case the increment on r_{att} is bounded by 0.001p.u., which is almost indistinguishable from the impact of measurement noises. The strong hiddenness of EXR-MTD in power distribution systems is due to the small branch ratios X/R, under which the reactance perturbation merely has neglectable impact on the branch impedance.

5.5 Summary

In this chapter, we attempted to conduct an explicit analysis on the MTD performance in AC power systems based on measurement residuals. To handle the nonlinear dynamics, we adopted the sensitivity analysis around the optimum point and derived explicit approximations of residuals, with which we designed EXR-MTD to jointly optimize the effectiveness and hiddenness. Through simulation results, EXR-MTD was shown to have the stronger hiddenness than existing MTD strategies, while the effectiveness is stronger than or comparable to those of existing MTD strategies. It was observed that the impact of EXR-MTD on the OPF generation cost is negligible as the strong hiddenness limits the fluctuations of power flows, and the computation time of EXR-MTD is tolerable even for the polish system test cases that includes more than 2000 buses, both indicating that EXR-MTD has the potential to be applied in real-world power systems. Furthermore, the EXR-MTD in power distribution systems possesses natural hiddenness as the branches usually have small X/R ratios, which means that more metrics are required for the better design of MTD therein besides effectiveness and hiddenness.

References

1. T. Seals, Ransomware attacks hit major utilities. [EB/OL], https://threatpost.com/ransomware-attacks-major-utilities/163687/. Accessed 23 Jul 2021
2. P. Dobson, Western venezuela rocked by rolling blackouts. [EB/OL], https://venezuelanalysis.com/news/14873/. Accessed 23 Jul 2021

3. Y. Liu, P. Ning, M.K. Reiter, False data injection attacks against state estimation in electric power grids. ACM Trans. Inf. Syst. Secur. **14**(1), 13 (2011)
4. R. Deng, P. Zhuang, H. Liang, False data injection attacks against state estimation in power distribution systems. IEEE Trans. Smart Grid **10**(3), 2871–2881 (2019)
5. X. Liu, Z. Li, False data attacks against AC state estimation with incomplete network information. IEEE Trans. Smart Grid **8**(5), 2239–2248 (2016)
6. Z. Zhang, R. Deng, P. Cheng, M.-Y. Chow, Strategic protection against FDI attacks with moving target defense in power grids. IEEE Trans. Control Netw. Syst. 1–1 (2021)
7. M. Liu, C. Zhao, Z. Zhang, R. Deng, P. Cheng, J. Chen, Converter-based moving target defense against deception attacks in DC microgrids. IEEE Trans. Smart Grid 1–1 (2021)
8. D. Divan, H. Johal, Distributed FACTS-A new concept for realizing grid power flow control, in *IEEE PESC*, (IEEE, 2005), pp. 8–14
9. Z. Zhang, R. Deng, D.K.Y. Yau, P. Cheng, J. Chen, Analysis of moving target defense against false data injection attacks on power grid. IEEE Trans. Inf. Forensics Secur. **15**, 2320–2335 (2020)
10. B. Li, G. Xiao, R. Lu, R. Deng, H. Bao, On feasibility and limitations of detecting false data injection attacks on power grid state estimation using d-facts devices. IEEE Trans. Ind. Inform. **16**(2), 854–864 (2019)
11. C. Liu, J. Wu, C. Long, D. Kundur, Reactance perturbation for detecting and identifying FDI attacks in power system state estimation. IEEE J. Sel. Top. Signal Process. **12**(4), 763–776 (2018)
12. B. Liu, H. Wu, Optimal D-FACTS placement in moving target defense against false data injection attacks. IEEE Trans. Smart Grid **11**(5), 4345–4357 (2020)
13. S. Lakshminarayana, D.K. Yau, Cost-benefit analysis of moving-target defense in power grids. IEEE Trans. Power Syst. **36**(2), 1152–1163 (2021)
14. J. Tian, R. Tan, X. Guan, T. Liu, Enhanced hidden moving target defense in smart grids. IEEE Trans. Smart Grid **10**(2), 2208–2223 (2018)
15. I. Markwood, Y. Liu, K. Kwiat, C. Kamhoua, Electric grid power flow model camouflage against topology leaking attacks, in *IEEE INFOCOM*, (IEEE, 2017), pp. 1–9
16. J. Kim, L. Tong, R.J. Thomas, Subspace methods for data attack on state estimation: a data driven approach. IEEE Trans. Signal Process. **63**(5), 1102–1114 (2014)
17. Z. Zhang, R. Deng, D.K. Yau, P. Cheng, J. Chen, On hiddenness of moving target defense against false data injection attacks on power grid. ACM Trans. Cyber Phys. Syst. **4**(3), 1–29 (2020)
18. B. Liu, H. Wu, Optimal planning and operation of hidden moving target defense for maximal detection effectiveness. IEEE Trans. Smart Grid **12**(5), 4447–4459 (2021)
19. C. Liu, H. Liang, T. Chen, J. Wu, C. Long, Joint admittance perturbation and meter protection for mitigating stealthy FDI attacks against power system state estimation. IEEE Trans. Power Syst. **35**(2), 1468–1478 (2019)
20. M. Cui, J. Wang, Deeply hidden moving-target-defense for cybersecure unbalanced distribution systems considering voltage stability. IEEE Trans. Power Syst. **36**(3), 1961–1972 (2021)
21. B. Liu, H. Wu, A. Pahwa, F. Ding, E. Ibrahim, T. Liu, Hidden moving target defense against false data injection in distribution network reconfiguration, in *IEEE PESGM*, pp. 1–5 (2018)
22. M. Liu, C. Zhao, Z. Zhang, R. Deng, and P. Cheng, Analysis of moving target defense in unbalanced and multiphase distribution systems considering voltage stability, in *SmartGridComm*, pp. 207–213 (2021)
23. A. Monticelli, *State Estimation in Electric Power Systems: a Generalized Approach* (Springer Science & Business Media, 2012)
24. P. Zhuang, R. Deng, H. Liang, False data injection attacks against state estimation in multiphase and unbalanced smart distribution systems. IEEE Trans. Smart Grid **10**(6), 6000–6013 (2019)
25. C. Rakpenthai, S. Premrudeepreechacharn, S. Uatrongjit, N.R. Watson, An optimal PMU placement method against measurement loss and branch outage. IEEE Trans. Power Deliv. **22**(1), 101–107 (2007)

26. C. Muscas, M. Pau, P.A. Pegoraro, S. Sulis, Effects of measurements and pseudomeasurements correlation in distribution system state estimation. IEEE Trans. Instrum. Meas. **63**(12), 2813–2823 (2014)
27. A. Primadianto, C.-N. Lu, A review on distribution system state estimation. IEEE Trans. Power Syst. **32**(5), 3875–3883 (2016)
28. K. Madsen, H.B. Nielsen, O. Tingleff, Methods for non-linear least squares problems (2004)
29. J. Allemong, L. Radu, A. Sasson, A fast and reliable state estimation algorithm for AEP's new control center. IEEE Trans. Power Appar. Syst. **4**, 933–944 (1982)
30. A. Abur, A.G. Exposito, *Power System State Estimation: Theory and Implementation* (CRC Press, 2004)
31. NESCOR, Electric sector failure scenarios and impact analyses version 3.0 (2020), [Online]. https://smartgrid.epri.com/doc/NESCOR%20Failure%20Scenarios%20v3%2012-11-15.pdf
32. E. Castillo, J.M. Gutiérrez, A.S. Hadi, Sensitivity analysis in discrete Bayesian networks. IEEE Trans. Syst. Man Cybern. Part A Syst. Hum. **27**(4), 412–423 (1997)
33. P.A. Garcia, J.L.R. Pereira, S. Carneiro, V.M. Da Costa, N. Martins, Three-phase power flow calculations using the current injection method. IEEE Trans. Power Syst. **15**(2), 508–514 (2000)
34. S. Chatterjee, A.S. Hadi, *Sensitivity Analysis in Linear Regression*, vol. 327 (John Wiley & Sons, 2009)
35. Z. Li, S. He, S. Zhang, *Approximation Methods for Polynomial Optimization: Models, Algorithms, and Applications* (Springer Science & Business Media, 2012)

Chapter 6
Impact Analysis of Moving Target Defense in the Smart Grid

6.1 Background

6.1.1 Small Signal Stability

In the power system, the small signal model consists of the rotor and stator dynamics of the synchronous generator [12]. Using the linearization approach [13], the small signal stability model can be written as

$$\Delta \dot{x} = A\Delta x + I\Delta u, \tag{6.1}$$

where Δ means the change of the variable, $x \in \Re^N$ is the state vector, $u \in \Re^N$ is the input vector, $A \in \Re^{N\times N}$ is the system matrix, and I is an identity matrix. The small signal stability is determined by the eigenvalues of matrix A. If the real parts of the eigenvalues of A are all negative, the system is stable; otherwise, the system is unstable. The spectral abscissa of A is the maximum real part of its eigenvalues, which is

$$\tilde{\sigma}\,(A) = \max\{real\,(\lambda_i) : \det\,(A - \lambda I) = 0\}, \tag{6.2}$$

where $real(\cdot)$ is the real part of the variable, $\det(\cdot)$ computes the determinant of an matrix, λ is a vector of eigenvalues, and $1 \le i \le N$. Therefore, if $\tilde{\sigma}\,(A) < 0$, then the system is stable; otherwise, the system is unstable.

As shown in Fig. 6.1, the single-machine infinite bus model is

$$\begin{aligned} \frac{\partial \delta}{\partial t} &= \omega - \omega_s, \\ M\frac{\partial \omega}{\partial t} &= T_m - T_e - D(\omega - \omega_s), \end{aligned} \tag{6.3}$$

R. Deng et al., *Moving Target Defense in the Smart Grid*, Wireless Networks,
https://doi.org/10.1007/978-3-031-92378-4_6

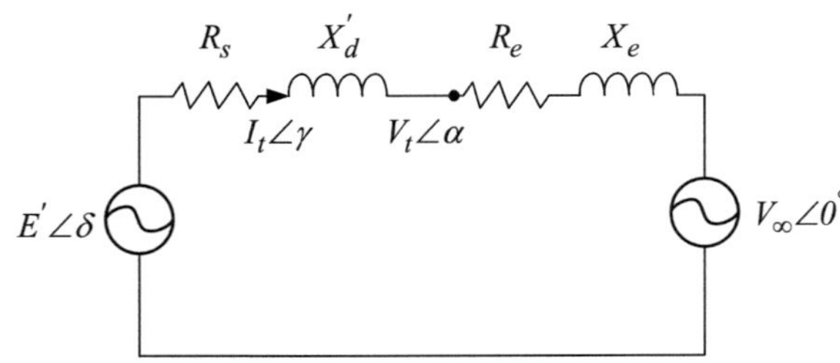

Fig. 6.1 The single-machine infinite bus system

where δ is the phase angle of the generator, ω is the generator angular speed, ω_s is the synchronous speed, $M = \frac{2H}{\omega_s}$, H is the inertial constant, T_m is the input mechanical power, T_e is the active electric power delivered by the generator to the infinite bus, D is the damping constant, and ω_s is the reference angular speed.

The electromagnetic torque T_e is

$$T_e = \frac{E' V_\infty \sin\delta}{X_d' + X_e}, \tag{6.4}$$

where E' is the voltage of the generator, V_∞ is the voltage of the infinite bus, and X_e is the equal reactance. Since resistance is much smaller than reactance, resistance is ignored here. By linearizing at the stable operating point, we can get

$$\begin{aligned} \frac{\partial \Delta\delta}{\partial t} &= \Delta\omega, \\ M\frac{\partial \Delta\omega}{\partial t} &= -\Delta T_e - D\Delta\omega. \end{aligned} \tag{6.5}$$

Further, we obtain that

$$\Delta T_e = \frac{E' V_\infty \cos\delta_0}{X_d' + X_e}\Delta\delta = K\Delta\delta, \tag{6.6}$$

where $K = \frac{E' V_\infty \cos\delta_0}{X_d' + X_e}$. Rewriting Eq. (6.5) into the matrix form, we get

$$\begin{bmatrix} \Delta\dot{\delta} \\ \Delta\dot{\omega} \end{bmatrix} = \begin{bmatrix} 0 & 1 \\ -\frac{1}{M}K & -\frac{1}{M}D \end{bmatrix} \begin{bmatrix} \Delta\delta \\ \Delta\omega \end{bmatrix}. \tag{6.7}$$

Therefore, from (6.1), $\Delta x = \begin{bmatrix} \Delta\delta \\ \Delta\omega \end{bmatrix}$, and the matrix A is

$$\boldsymbol{A} = \begin{bmatrix} 0 & 1 \\ -\frac{1}{M}K & -\frac{1}{M}D \end{bmatrix}. \tag{6.8}$$

6.1.2 Load Redistribution Attack

However, the power system is vulnerable to cyberattacks because the information and communication infrastructure has plenty of unnoticed entry points for attackers [11, 20]. As shown in Fig. 6.2, the attackers can construct stealthy attacks on the load measurements. A typical cyberattack is the LRA.

The LRA keeps the sum of loads unchanged after manipulating some vulnerable load measurements. That is, the loads of some buses are increased while the loads of other buses are reduced. Suppose the load power vector P_d is affected by the malicious perturbations ΔP_d^a, then the modified load vector is

$$\begin{aligned} &P_d^a = P_d + \Delta P_d^a, \\ &-\epsilon P_d \leq \Delta P_d^a \leq \epsilon P_d, \\ &\sum_{i \in \mathcal{V}_d} \Delta P_{d,i}^a = 0, \end{aligned} \tag{6.9}$$

where ϵ indicates the percentage of the injected measurement error, $\Delta P_{d,i}^a$ is the ith element of the load measurement vector ΔP_d^a, and $\mathcal{V}_d$ is the set of attacked load. To realize a stealthy LRA, the injected load perturbation ΔP_d^a should be within a reasonable bound to avoid detection. However, the modification of load measurements also leads to the change of the system state x, power injection measurement P, and power flow measurement P_f. Overall, we have

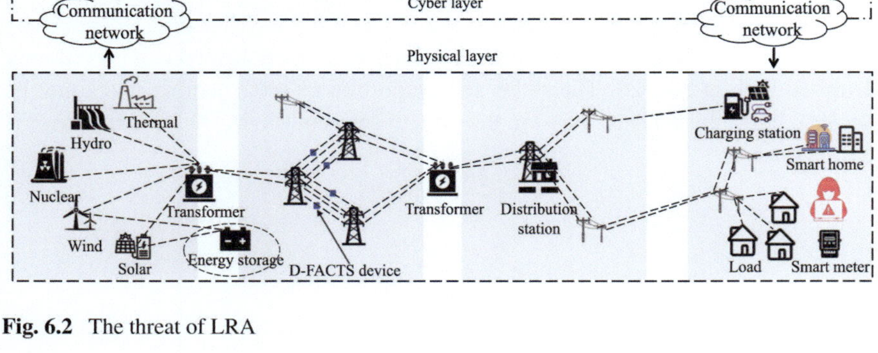

Fig. 6.2 The threat of LRA

$$\begin{aligned} \Delta P^a &= K_d \Delta P_d^a = U \Delta x, \\ \Delta P_f^a &= K_s K_d \Delta P_d^a = S \Delta x, \end{aligned} \tag{6.10}$$

where U is the branch-branch susceptance matrix, S is bus-branch shift factor matrix, K_d is the bus-load incidence matrix, and K_s is the injection shift factor matrix. With the DC model [4], we have

$$[P, P_f, -P_f]^T = [U, S, -S]^T x + e. \tag{6.11}$$

If $z_a = [P^a, P_f^a, -P_f^a]^T = z + a$, where $a = [\Delta P^a, \Delta P_f^a, -\Delta P_f^a]^T$, we have $a = Hc$, which can bypass BDD [10].

6.1.3 Moving Target Defense

The phyMTD approach is a well-known defense approach realized by perturbing the physical parameters of the power system [5, 17]. As given in [17], the D-FACTS device is used to perturb the branch reactance for detecting FDI attacks. Suppose the reactance of branch j is perturbed, then we have

$$X_j \overset{\text{phyMTD}}{\longrightarrow} \tilde{X}_j, \tag{6.12}$$

where X_j is the reactance of branch j, $\tilde{X}_j$ is branch reactance after phyMTD, and $\overset{\text{phyMTD}}{\longrightarrow}$ means the change of variable after phyMTD. As a result, the measurement matrix H is changed to $\tilde{H}$ after phyMTD. Many studies use the rank of the composite matrix $[H\ \tilde{H}]$ to measure the phyMTD's effectiveness [6, 14]. However, the rank-based method is not accurate and scalable, which has been proved in [18]. This forces us to propose a new indicator to quantify the effectiveness of phyMTD. Since most systems have buses or super buses with degree 1, phyMTD cannot achieve completeness [3].

Inspired by the conclusion obtained in [19], we observe that the bus cannot be attacked if more than one cyclic branch[1] (defined in [19]) connected to it is randomly perturbed with D-FACTS. Therefore, the proportion of protected buses (cannot be attacked) after phyMTD to measure the effectiveness of phyMTD. This indicator is described as

$$\varphi = \frac{|B|}{m} \times 100\%, \tag{6.13}$$

[1] The cyclic branch means it can participate in forming a circle with other branches.

where m is the number of buses, B denotes the set of protected buses after phyMTD, and $|\cdot|$ computes the size of the set. The set B is closely related to the selection of perturbed cyclic branches Ω. The set B is composed of buses connected by perturbed cyclic branches in the set Ω. If the set Ω of perturbed cyclic branches is determined, the set B of protected buses can be obtained correspondingly.

Note that if perturb branches randomly, only a few buses may be protected, which means that the effectiveness of phyMTD cannot be guaranteed and D-FACTS device resources will be wasted. Assume the number of D-FACTS devices is the same. When the branch perturbation is randomly set and the perturbed branches only contain a few cyclic branches. It results in that only a few buses can be protected, that is, the indicator φ is small. When branches are selected from the cyclic branches set for installation and perturbation, in this case, not only the indicator φ is the largest, the number of D-FACTS devices required is also the least. Therefore, by optimizing the number and installation location of D-FACTS devices, the device cost can be minimized and the defense effect can be maximized. Therefore, finding the cyclic branches is very important for the deployment of phyMTD. Therefore, finding the cyclic branches is very important for the deployment of phyMTD. Thus we propose Algorithm 1 to find all cyclic branches in the power transmission network.

Then, choose a suitable location to deploy the D-FACTS devices, that is, select the perturbed cyclic branches set Ω, thus obtain the protected buses set B, and further obtain the indicator φ. The main idea of this algorithm is based on the recursive depth-first search (DFS) algorithm, which uses DFS to obtain the set Ω_c of all cyclic branches in the system. Note that the indicator φ can adapt to topology changes, its calculation can be completed in linear time. It is sufficient to perform a depth-first traversal again on the buses connected to the added or deleted branches, and then update Ω_c based on the results. Hence, the computation of this indicator does not affect the real-time monitoring of the power system security.

Here we briefly introduce the algorithm. First, the bus-branch information matrix G can be obtained through the topology information of the power transmission network. Here, the matrix G is similar to the adjacency matrix of the system, the difference is that the non-zero elements in the matrix G denote the index of the branch. Second, the DFS function is called to traverse each bus and find all bus sequences that form a circle in the power transmission network. When a circle is found, the cyclic branches can be determined according to the corresponding bus sequence, and they are saved in the set Ω_c. This is because that the branch connected to the previous and next buses in the bus sequence is the cyclic branch. Finally, after traversing all buses, the set Ω_c can be obtained. In Algorithm 4, $path$ denotes the bus sequence on the current DFS path, $neighbor$ denotes the neighbor bus of the current bus, and seq denotes the found bus sequence that can form a circle.

Algorithm 4: Algorithm for computing the set Ω_c

```
Input: The bus-branch information matrix G;
Output: Ω_c
Initialize Ω_c = {}
for each bus in G do
    DFS(bus, Ω_c, []);
end
Function DFS(bus, Ω_c, path):
    path = [path bus];
    for each unvisited neighbor of bus do
        if neighbor in path then
            seq = path[path.index(neighbor), :];
            for i = 1:size(seq) do
                if i ≠ size(seq) then
                    Ω_c = {G[seq(i), seq(i + 1)]} ∪ Ω_c
                else
                    Ω_c = {G[seq(size(seq)), seq(1)]} ∪ Ω_c
                end
            end
        else
            DFS(neighbors, Ω_c, path);
        end
    end
```

6.2 Problem Statement

Considering the small signal stability, the eigenvalue sensitivity of λ_i with respect to the branch parameter X is approximated by [16]

$$\frac{\partial \lambda_i}{\partial X} = \psi_i^T \frac{\partial A}{\partial X}\phi_i, \text{ for } 1 \leq i \leq N, \tag{6.14}$$

where ψ_i and ϕ_i are the left and right eigenvectors of the eigenvalue λ_i, respectively. The damping ratio is also an indicator of the operation performance of the system, which is computed by

$$\xi_i = \frac{-\alpha_i}{\sqrt{{\alpha_i}^2 + {\beta_i}^2}}, \text{ for } 1 \leq i \leq N, \tag{6.15}$$

where α_i and β_i are the real and imagine parts of the eigenvalue λ_i, respectively. If the damping ratio is small, the system suffers from oscillation, and the negative damping ratio will make the system unstable. The poorly damped (damping ratio less than 10%) or unstable modes are called electromechanical modes. The electromechanical modes can be used to measure the stability margin.

The participation matrix is used to evaluate the association between state variables and electromechanical modes [9].

The participation matrix Q is computed as

$$Q = \left[\, Q_1 \; Q_2 \cdots Q_n \,\right], \tag{6.16}$$

where

$$Q_i = \begin{bmatrix} Q_{1i} \\ Q_{2i} \\ \vdots \\ Q_{ni} \end{bmatrix} = \begin{bmatrix} \phi_{1i}\psi_{1i} \\ \phi_{2i}\psi_{2i} \\ \vdots \\ \phi_{ni}\psi_{ni} \end{bmatrix}, \tag{6.17}$$

where ϕ_{ki} is the kth element of the right eigenvector of the ith eigenvalue, ψ_{ki} is the kth element of the left eigenvector of the ith eigenvalue, and Q_i is the ith column of the participation matrix, representing a measure of the participation of the state variable in the ith eigenvalue.

In this chapter, we aim to derive the impact of the reactance perturbation (i.e., phyMTD) on the small signal stability and stability margin. From (6.8), the matrix A is affected by the matrices K, M, and D. However, K is related to the branch reactance X_e, which is an equivalent value of the overall reactance of the power system. Therefore, the perturbations of branch reactances change the eigenvalues of A. Since the phyMTD is realized by perturbing the branch reactances, the small signal stability is affected by the phyMTD.

With the New England 39-bus power system model, we validate the impact of phyMTD on the small signal stability. An arbitrary set of branches is selected and the corresponding reactances are increased by 20%. The results are shown in Fig. 6.3. After the perturbation, some eigenvalues move to the right half plane, indicating that the small signal stability is destroyed. Therefore, the inappropriate perturbations of branch reactances can destroy the small signal stability. More detailed explanations are given in Sect. 6.4.2.

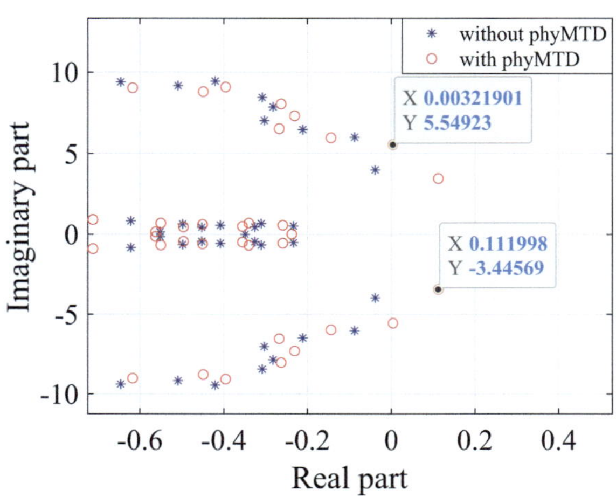

Fig. 6.3 The changes of eigenvalues with and without phyMTD

6.3 Small Signal Stability-Guaranteed Moving Target Defense

In this section, firstly, we propose the condition to guarantee the effectiveness of phyMTD and the small signal stability simultaneously. Secondly, we propose a constrained optimization problem to compute the bound of the reactance perturbation to realize the condition. Thirdly, we improve the small signal stability margin by optimizing the implementation of phyMTD.

6.3.1 *Condition for Realizing Small Signal Stability-Guaranteed Moving Target Defense*

To guarantee that the small signal stability is not breached, the real parts of all eigenvalues of matrix $\tilde{A}$ (i.e., $A \stackrel{\text{phyMTD}}{\longrightarrow} \tilde{A}$) after phyMTD must be negative, that is

$$\lambda_i \stackrel{\text{phyMTD}}{\longrightarrow} \tilde{\lambda}_i, real\left(\tilde{\lambda}_i\right) < 0, \text{ for } 1 \leq i \leq N, \tag{6.18}$$

where $\tilde{\lambda}_i$ is the ith eigenvalue of $\tilde{A}$ after phyMTD. The change of eigenvalue is related to the branches whose reactances are perturbed and how much they are perturbed. That is, small signal stability is closely related to the realization of phyMTD.

On the other hand, to guarantee the effectiveness of phyMTD, the following constraint must be satisfied, i.e.,

$$\varphi \geq \tau, \tag{6.19}$$

where τ is the threshold to ensure the effectiveness of phyMTD. Therefore, the effectiveness of phyMTD and the small signal stability of the system can be simultaneously guaranteed if the following two conditions are satisfied: (i) $\varphi \geq \tau$; (ii) $real\left(\tilde{\lambda}_i\right) < 0$, for $1 \leq i \leq N$.

6.3.2 *Bounded Reactance Perturbations*

Here we develop a method to calculate the perturbation bound of the branch reactance for guaranteeing the small signal stability. Based on the eigenvalue sensitivity with respect to the reactances, we have

$$\Delta\lambda_i = \frac{\partial \lambda_i}{\partial X}\Delta X = \psi_i^T \frac{\partial A}{\partial X}\phi_i \Delta X, \text{ for } 1 \leq i \leq N, \tag{6.20}$$

where $\Delta\lambda_i$ is the change of the ith eigenvalue of $\tilde{A}$ after phyMTD. Further, we have

$$\widetilde{\lambda_i} = \lambda_i + \Delta\lambda_i = \lambda_i + \sum_{j\in\Omega} \frac{\partial\lambda_i\left(X_j\right)}{\partial X_j}\Delta X_j, \text{ for } 1 \le i \le N, \tag{6.21}$$

where Ω is the set of branches whose reactances are perturbed after phyMTD, ΔX_j denotes the magnitude of the reactance perturbation of the jth branch. We can see that the small signal stability of the system is closely related to Ω and the perturbation size ΔX, which is a vector containing the reactance perturbations. However, to guarantee the small signal stability, the perturbations of branch reactances should be limited and bounded. Hence, we formulate an optimization problem to compute the set Ω and ΔX, that is,

$$\min_{\Delta X,\Omega} \quad ||\Delta X||_2 + \frac{|\Omega|}{l} \tag{6.22}$$

$$s.t. \quad \widetilde{\lambda_i} = real(\lambda_i) + \sum_{j\in\Omega} real(e_{ij})\Delta X_j, 1 \le i \le N, \tag{6.22-1}$$

$$y = \max\{\widetilde{\lambda}_1, \widetilde{\lambda}_2, \cdots, \widetilde{\lambda}_N\} \ge 0, \tag{6.22-2}$$

$$aX_{0j} \le \Delta X_j \le bX_{0j}, j \in \Omega \subseteq \Omega_c, \tag{6.22-3}$$

$$\varphi \ge \tau, \tag{6.22-4}$$

where $e_{ij} = \frac{\partial\lambda_i(X_j)}{\partial X_j}$ denotes the sensitivity of the ith eigenvalue against the reactance perturbation of the jth branch, $real(\cdot)$ is the real part of the variable, Ω_c denotes the set of all cyclic branches in the system, l is the number of system branches, a and b are the minimum and maximum perturbation ratios (usually, $a = -20\%$, $b = 20\%$) [8, 15], and X_{0j} is the original reactance of branch j.

If we find the appropriate Ω and ΔX that satisfy all constraints, then the obtained phyMTD is effective and guarantees the small signal stability. At this time, the number of D-FACTS devices required is relatively small, thus further reducing hardware and maintenance costs. If the set Ω of perturbed branch reactances is known, the problem (6.22) becomes the linear programming problem. Hence, the solution space of ΔX is the space surrounded by N hyperplanes. Therefore, the above optimization problem is equivalent to solving the problem of the minimum distance from zero point (ΔX is 0 when not perturbed) to N hyperplanes. Suppose that the minimum distance projection point is ΔX^*.

By solving the above optimization problem, a set of perturbation values ΔX^* and the perturbation branch set Ω^* can be obtained. X_0 can be regarded as the coordinate value of the initial point. Based on the solution of the optimization problem, we redefine the perturbation bound. The perturbation bound of phyMTD is designed as

$$LB < \widetilde{X} < UB, \tag{6.23}$$

where $\widetilde{X}$ denotes the vector of branch reactances after phyMTD. In this chapter, the perturbation bound is defined as the stable interval (LB, UB), where $LB = X_0 - abs\,(\Delta X^*)$, $UB = X_0 + abs\,(\Delta X^*)$, and $abs(\cdot)$ is the absolute value function.

The key to solving the optimization problem is Ω. When Ω is determined, we can also determine ΔX. Since the set Ω_c is limited by the network structure, the selection of the set Ω is also limited. The set Ω that meets the phyMTD's effectiveness corresponds to a set of ΔX is the optimization goal of the problem. However, as the system size increases, the number of feasible sets Ω increases exponentially. This is a computation-intensive problem. The number of sets Ω is $\sum_{i=1}^{N_c} C_{N_c}^i$, where N_c denotes the number of all cyclic branches. Therefore, we use a fireworks algorithm to solve the above optimization problem. To facilitate calculation, we regard the selection of the set Ω as the selection of the boolean vector $\pi \in \{0, 1\}^l$. Therefore, the fitness function is

$$f(\pi) = \min\{\frac{abs(-\lambda_i)}{\sqrt{\sum_{j=1}^{l} \pi_j e_{ij}}} \mid 1 \leq i \leq N\} + \frac{\sum_{j=1}^{l} \pi_j}{l}, \tag{6.24}$$

where $f(\pi)$ denotes the fitness when selecting the branch set π.

Besides, for π, the perturbation number r (explosion radius) and the number of fireworks num (the number of explosion sparks) are

$$r = \hat{r} \times \frac{f(\pi) - y_{min} + \varepsilon}{\sum_{i=1}^{num} (f(\pi_i) - y_{min} + \varepsilon)}, \tag{6.25}$$

$$num = \hat{num} \times \frac{y_{max} - f(\pi) + \varepsilon}{\sum_{i=1}^{num} (y_{max} - f(\pi_i)) + \varepsilon}, \tag{6.26}$$

where $y_{min} = \min(f(\pi))$ and $y_{max} = \max(f(\pi))$ is respectively the minimum and maximum value of the fitness in the current fireworks, $\hat{r}$ is a constant used to adjust the perturbation number, $\hat{num}$ is a constant used to adjust the number of fireworks in the next round, and ε is an exceptionally decreasing value to avoid the denominator going to zero. For example, when $\hat{\pi}$ corresponds to the next round of exploding fireworks, the selection of π for the next round should be based on $\hat{\pi}$, which is obtained by adding or subtracting a boolean vector according to the value $\bar{r}$, where $\bar{r} \in [-r, r]$. The π is updated as

$$\pi = \begin{cases} \hat{\pi} - \eta_1 & \bar{r} < 0 \\ \hat{\pi} & \bar{r} = 0, \\ \hat{\pi} + \eta_2 & \bar{r} > 0 \end{cases} \tag{6.27}$$

where η_1 and η_2 are boolean vectors and the number of 1s in them is $abs(\bar{r})$. The set corresponding to η_1 is Ω_{η_1}, satisfying $\Omega_{\eta_1} \in \Omega_{\hat{\pi}}$. The notation $\Omega_{\hat{\pi}}$ denotes the set corresponding to $\hat{\pi}$. The set corresponding to η_2 is Ω_{η_2}, satisfying $\Omega_{\eta_2} \in (\Omega_c - \Omega_{\hat{\pi}})$.

The result $\Omega_c - \Omega_{\hat{\pi}}$ denotes the unselected branch set. When π is selected in the next round, the perturbation number $\bar{r}$ is randomly generated. The boolean vector η with $\bar{r}$ elements of 1 is generated. Hence, π is generated according to Eq. (6.24). That is, $\hat{\pi}$ is used as a constraint to limit the random selection of π in the next round.

The critical steps of solving the problem (6.22) are as follows (an illustration is given in Fig. 6.4):

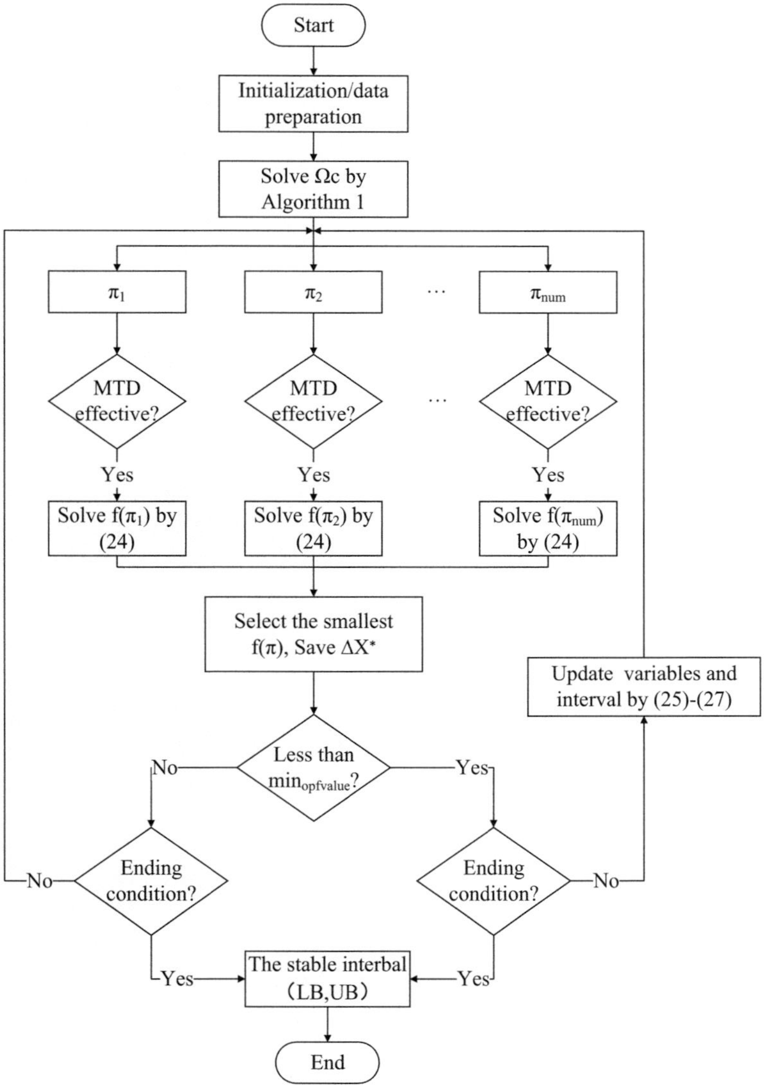

Fig. 6.4 The process of solving the optimization problem (6.22) using the fireworks algorithm

- Step 1: The bus-branch information matrix G is obtained according to the topology information of the power system, which is used to determine the set Ω_c through Algorithm 4. We initialize the variables num, $\hat{num}$, $\hat{r}$, and ε.
- Step 2: The optimization problem is solved through the fireworks algorithm. First, we randomly select num sets Ω (i.e., π) from the set Ω_c, expressed as $\{\pi_1\ \pi_2\ \ldots\ \pi_{num}\}$. Second, if π satisfies the constraint $\varphi \geq \tau$, the fitness $f(\pi)$ can be solved through Eq. (6.24). Further, the π^* (i.e., Ω^*) with the smallest fitness is selected as the basis for the next round of updates. The minimum ΔX^* is calculated under the condition of π^* using the linear programming problem. In addition, to avoid falling into local optimality, we also randomly selected a π from the remaining set as the basis for the next round of updates. Then, we determine the minimum fitness y_{min} whether it is less than $\min_{opfvalue}$ or not. The initial value of $\min_{opfvalue}$ is given arbitrarily. If it is less than $\min_{opfvalue}$, the number of fireworks in the next round num and the perturbation number r need to be updated through Eqs. (6.25)–(6.26). Moreover, the π of the next round is generated according to Eq. (6.27). Note that the optimal π^* and the randomly selected π are used separately to update the variables and the stable interval in the next round. Otherwise, the involved variables are not updated. Finally, the optimal solutions of Ω^* (i.e., π^*) and ΔX^* are obtained, which might be suboptimal.
- Step 3: The value of ΔX^* is used to solve $LB = X_0 - abs\,(\Delta X^*)$, $UB = X_0 + abs\,(\Delta X^*)$ and obtain the stable interval (LB, UB).

With the set Ω^*, if the branch reactance is perturbed within the stable interval (LB, UB), the small signal stability of the system and the effectiveness of phyMTD are guaranteed simultaneously. Besides, in our opinion, the hardware and maintenance costs are proportional to the number of D-FACTS devices. Therefore, in the optimization problem (6.22), the second part is used to minimize the number of perturbed branches, that is, to minimize the number of the deployed D-FACTS devices, thereby minimizing the cost.

In addition, we analyze the limitations of the proposed method. First, the local optimal point might be obtained since we use an intelligent algorithm to solve the optimization problem. Note that a local optimum is sufficient to construct a phyMTD that can guarantee both the effectiveness of defending against stealthy attacks and the small signal stability. Second, the time complexity of the proposed method is linearly related to the system scale. Among them, the firework algorithm consumes the most time, which is approximately $O(n \times m \times d)$, where n is the number of fireworks, m is the number of sparks generated, and d is the dimension of the fireworks (the number of cyclic branches). Finally, the space complexity of the proposed method is linearly related to the system size. This is because the proposed method only involves the storage of system data and additional cyclic branches, fitness values, and the optimal solution.

6.3.3 Optimization of Stability Margin

Further, we improve the margin of the small signal stability by optimizing the implementation of phyMTD. Here, we take the minimum damping ratio as the stability margin. It can be seen from Sect. 6.1 and Eq. (6.14) that the damping ratio is affected by the perturbations of the branch reactances. Since $\lambda = \alpha + j\beta$, we have

$$\frac{\partial \lambda_i}{\partial X} = \frac{\partial \alpha_i}{\partial X} + j\frac{\partial \beta_i}{\partial X} = \psi_i^T \frac{\partial A}{\partial X}\phi_i, \tag{6.28}$$

where

$$\frac{\partial \alpha_i}{\partial X} = real\left(\frac{\partial \lambda_i}{\partial X}\right) = real\left(\psi_i^T \frac{\partial A}{\partial X}\phi_i\right), \tag{6.29}$$

$$\frac{\partial \beta_i}{\partial X} = imag\left(\frac{\partial \lambda_i}{\partial X}\right) = imag\left(\psi_i^T \frac{\partial A}{\partial X}\phi_i\right), \tag{6.30}$$

where $imag(\cdot)$ computes the imaginary part. Based on Eq. (6.15), we can derive the sensitivity of the damping ratio with respect to the reactance perturbation as

$$\begin{aligned} \frac{\partial \xi_i}{\partial X} &= \frac{\partial \xi_i}{\partial \alpha_i}\frac{\partial \alpha_i}{\partial X} + \frac{\partial \xi_i}{\partial \beta_i}\frac{\partial \beta_i}{\partial X} \\ &= \left[-\alpha_i^{-3}\beta_i^2\left(1+\frac{\beta_i^2}{\alpha_i^2}\right)^{-\frac{3}{2}}\right]\frac{\partial \alpha_i}{\partial X} + \left[\alpha_i^{-2}\beta_i\left(1+\frac{\beta_i^2}{\alpha_i^2}\right)^{-\frac{3}{2}}\right]\frac{\partial \beta_i}{\partial X}. \end{aligned} \tag{6.31}$$

Substituting Eqs. (6.29) and (6.30) into (6.31), we can get the sensitivity of the damping ratio with respect to the reactance perturbation. Therefore, the change of damping ratio after phyMTD is described as

$$\Delta \xi_i = \frac{\partial \xi_i}{\partial X}\Delta X, \text{ for } 1 \le i \le N, \tag{6.32}$$

where $\Delta \xi_i$ is the change of ξ_i after phyMTD. Therefore, the damping ratio $\widetilde{\xi_i}$ after phyMTD is

$$\widetilde{\xi_i} = \xi_i + \Delta \xi_i = \xi_i + \sum_{j\in\Omega} \frac{\partial \xi_i\left(X_j\right)}{\partial X_j}\Delta X_j, \text{ for } 1 \le i \le N. \tag{6.33}$$

Further, to improve the margin of the small signal stability, the optimization problem is formulated as

$$\begin{aligned}
&\min_{\Delta X,\Omega} \quad ||1-\widetilde{\xi}||_\infty + \frac{|\Omega|}{l} \\
&s.t. \;\; \widetilde{\xi}_i = \xi_i + \sum_{j\in\Omega} \tfrac{\partial \xi_i(X_j)}{\partial X_j}\Delta X_j > 0, \text{ for } 1 \le i \le N \\
&\qquad aX_{0j} \le \Delta X_j \le bX_{0j}, \text{ for } j \in \Omega \subseteq \Omega_c \\
&\qquad \varphi \ge \tau,
\end{aligned} \tag{6.34}$$

where $||\cdot||_\infty$ is the infinite norm, which is used to improve the stability margin as much as possible.

The above problem is solved as follows. If the set Ω is known, the above optimization problem can be regarded as a combinatorial optimization problem. Since ΔX is constrained according to the perturbation number, its solution space is limited, there must be an optimal solution. However, the selection of the set Ω can also be regarded as a combinatorial optimization problem, which has been proved in Section V-B, making the solution to the above problem complicated. Therefore, a nested fireworks algorithm is proposed to solve the above problem. Here, the selection of the set Ω is also regarded as the selection of the boolean vector π. Then, the fitness equations for the outer and inner layer optimization problems are given. Note that the updated equations of the inner layer of perturbation number r_x and the number of fireworks num_x are similar to Eqs. (6.25)–(6.26), we are not repeating them. The fitness equation of the inner layer optimization problem is described as

$$g_1(\Delta X) = ||1-\widetilde{\xi}||_\infty, \tag{6.35}$$

where $g_1(\Delta X)$ denotes the fitness value corresponding to different ΔX under π. The fitness equation of the outer optimization problem is defined as

$$g_2(\pi) = \min_g + \frac{\sum_{j=1}^{l} \pi_j}{l}, \tag{6.36}$$

where $\min_g = \min(g_1(\Delta X))$ is the minimum value (optimal value) of fitness, which is solved by the inner firework algorithm, and $g_2(\pi)$ is the fitness value.

The critical steps of solving the problem (6.34) are as follows (an illustration is given in Fig. 6.5):

- Step 1: The bus-branch information matrix G is obtained according to the topology information of the power system, which is used to determine the set Ω_c through Algorithm 4. The variables num, $\hat{num}$, $\hat{r}$, ε, num_x, r_x, $\min_{opfvalue}$, and $\min_g$ are initialised.
- Step 2: The optimization problem is solved through the nested fireworks algorithm. First, the num sets Ω (i.e., π) are randomly selected from the set Ω_c, expressed as $\{\pi_1 \; \pi_2 \; \ldots \; \pi_{num}\}$. Second, when π satisfies the constraint $\varphi \ge \tau$, the fireworks algorithm is used again to generate num_x groups ΔX randomly. The corresponding ξ and fitness $g_1(\Delta X)$ are solved through Eq. (6.35). Then, we determine whether g_{1min} is less than $\min_g$ or not. If it is less than $\min_g$, $\min_g$ needs to be updated,

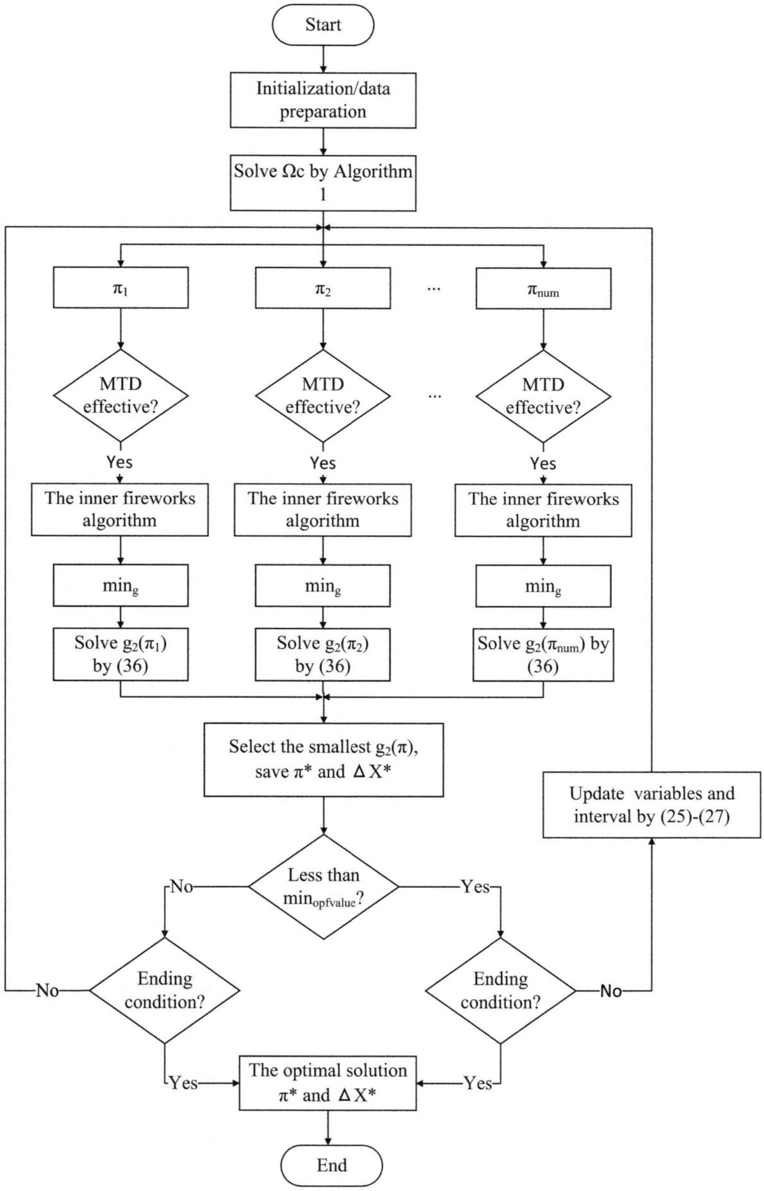

Fig. 6.5 The process of solving the optimization problem (6.34) using the nested fireworks algorithm

and the corresponding ΔX_{opf} is saved. The number num_x of fireworks in the next round and the perturbation number r_x are also updated, thereby updating the perturbation interval of ΔX, that is, $[\Delta X_{opf} - r_x \times \Delta X_{opf}, \Delta X_{opf} + r_x \times \Delta X_{opf}]$. Otherwise, these variables are not updated.

- Step 3: Then, the fitness $g_2(\pi)$ is solved under different π. The minimum fitness g_{2min} is decided whether it is less than $\min_{opfvalue}$ or not. If it is less than $\min_{opfvalue}$, the π^* corresponding to g_{2min} and ΔX^* are saved. Besides, the number num of fireworks in the next round and the perturbation number r are updated through Eqs. (6.25)–(6.26). The π is generated in the next round according to Eq. (6.27). Apart from the selection of the optimal π^*, another π is also randomly selected as the next round of explosive fireworks. Otherwise, there are no updates. After several iterations, the optimal solutions of Ω^* (i.e., π^*) and ΔX^* are obtained, which might be suboptimal.

When perturbing the branch reactances with the obtained Ω^* and ΔX^*, the stability margin of the system can be further improved. Meanwhile, the effectiveness of phyMTD is guaranteed.

6.4 Simulation Results

In this section, we evaluate the performance of the small signal stability-guaranteed MTD with the IEEE 39-bus test system and the IEEE 68-bus test system [7].

6.4.1 *Effectiveness of Moving Target Defense with Proposed Indicator*

To validate the MTD's effectiveness with the indicator φ, we compute the detection rate of LRAs with different φ. First, we construct 1000 LRA vectors according to Eqs. (6.9)–(6.11) as an attack pool. Then, under different branch perturbation settings, the corresponding LRA detection rates are obtained based on a Monte Carlo simulation executed 1000 times. The results are shown in Fig. 6.6. Since the 30-bus, 39-bus, and 68-bus power systems do not satisfy the completeness of phyMTD [14], therefore, the value of φ cannot reach 100%. We can see from Fig. 6.6 that the detection rate of LRAs increases with the value of φ. We also find that with the same φ, the larger the system size is, the higher the LRA detection rate. When the value of φ is larger than 22%, the LRA detection rate with these three systems exceeds 90%. Therefore, it is feasible to use $\varphi \geq \tau$ to measure the effectiveness of phyMTD. Taking the 68-bus power system as an example, when $\tau = 60\%$, the phyMTD is very effective as it almost can detect all LRAs in the attack pool.

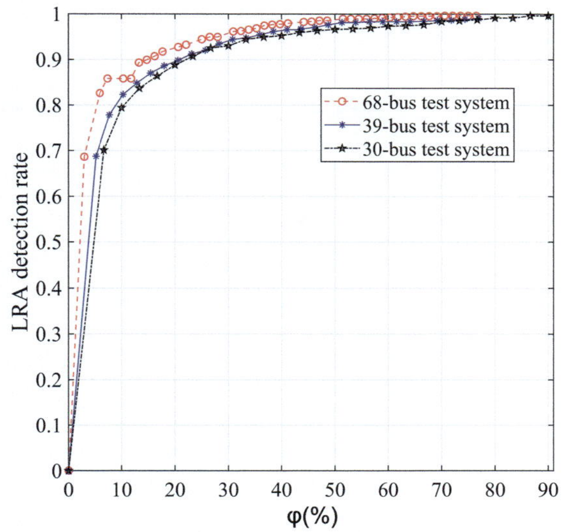

Fig. 6.6 The relationship between the indicator φ and LRA detection rate

6.4.2 *Impact of Moving Target Defense on Small Signal Stability*

Here we demonstrate that inappropriate reactance perturbations can destroy the small signal stability with the New England 39-bus and the IEEE 68-bus power systems.

To demonstrate our findings, we realize the phyMTD with the maximum composite rank and full-bus cover, respectively. Here, we define D_1 denotes the phyMTD with the maximum composite rank, and D_2 denotes the phyMTD with full deployment. The following cases are considered in this chapter. In case 1, we select a set of branches that can realize D_1 and randomly perturb the reactances of these branches within the limited perturbation limitation. In case 2, we select a set of branches that can realize D_1. The sign of the eigenvalue sensitivity is used as the perturbation direction and the branch reactances are randomly perturbed within the perturbation limitation. In case 3, we randomly perturb all branches within the perturbation limitation. In case 4, the sign of the eigenvalue sensitivity is used as the perturbation direction and we randomly perturb the reactances of all branches within the perturbation limitation.

(1) The New England 39-bus power system: First, the perturbation limitation of the branch reactance is set as $\pm 20\%$, as used in most studies [8, 15]. Second, assumed that each generator (except the reference bus) is equipped with automatic voltage regulation (AVR) and power system stabilizer (PSS) devices.

We conduct experiments with the above four cases, respectively. A quick conclusion is that the small signal stability is destroyed in all cases. The experimental results are shown in Fig. 6.7. We can see that case 1, case 2, case 3, and case 4 respectively have 4, 859, 23, and 1776 times to destroy the small signal stability. Furthermore, we

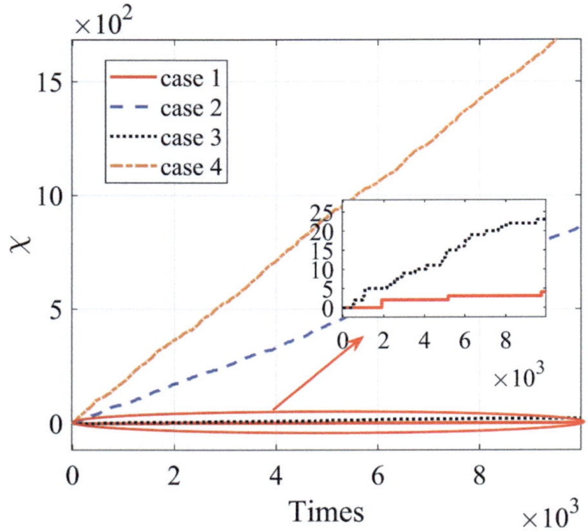

Fig. 6.7 The number of times the small signal stability of the system is destroyed under different cases. the notation χ denotes the number of times that the system is unstable

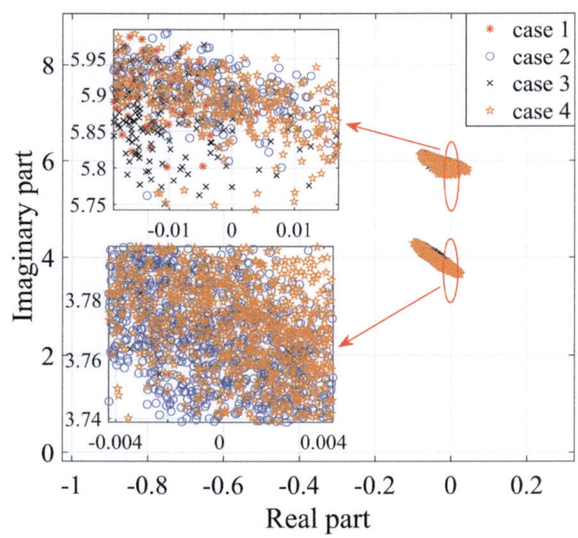

Fig. 6.8 The change trajectory of the spectral abscissa $\tilde{\sigma}(A)$ under different cases

analyze the trajectory of the spectral abscissa in these cases. As shown in Fig. 6.8, we can see that there are some spectral abscissas greater than 0, which indicates that small signal stability is breached.

(2) The IEEE 68-bus power system: Next, we use the IEEE 68-bus power system to validate our findings. Considering the cost, it is impossible to install the PSS devices at each generator in practice. Therefore, we assume that only five PSS devices are installed in the IEEE 68-bus power system. To find the optimal positions for deploying the PSS devices, we analyze the electromechanical mode of the system.

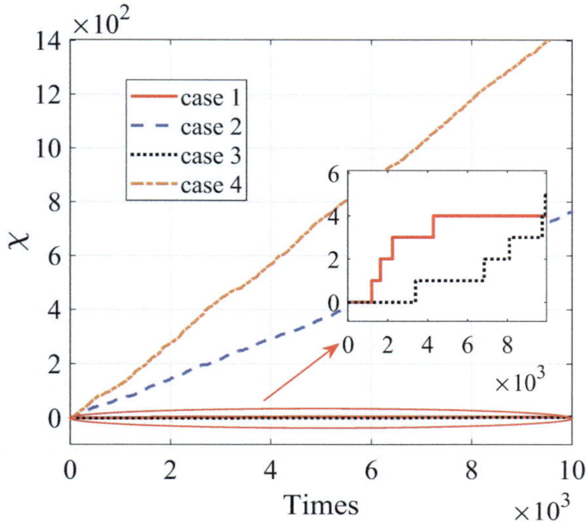

Fig. 6.9 The number of times the small signal stability of the system is destroyed under different cases. The notation χ denotes the number of times that the system is unstable

Table 6.3 shows the electromechanical modes with normalized participation factors. We can see that generators at buses 13, 9, 2, 14, 10, 12, 15, 5, 11, 8, 7, 4, 3, 6, 16, and 1 are the optimal choices for deploying PSS devices to increase the system's damping ratio in turn. This is because the above locations in turn have greater participation for smaller damping ratios. For example, as shown in Table 6.3, for the damping ratio -0.040791, the participation factor of bus 9 is the largest, so installing the PSS device at bus 9 can maximizes the improvement of this damping ratio. However, since the generators at buses 13–16 are virtual. The PSS devices cannot be installed on these buses. Therefore, we choose to install the PSS devices on generators at buses 2, 5, 9, 10, and 12.

Further, we analyze the impact of randomly perturbed branch reactances on the small signal stability within the perturbation limitation $\pm 20\%$. We consider D_1 and D_2, respectively. Then, we again conducted the experiment under the above four cases. The experimental results are shown in Fig. 6.9. We find that the phyMTD realized in all cases can destroy the small signal stability. Besides, we show the trajectory of the spectral abscissa after the phyMTD in Fig. 6.10. We can see that there are several spectral abscissas greater than 0 in all cases, which means the violation of the small signal stability.

6.4.3 Bounded Perturbation for Moving Target Defense

Here we validate the implementation of phyMTD to guarantee its effectiveness and its small signal stability simultaneously. To avoid the small signal stability from being destroyed by phyMTD, we solve the optimization problem (6.22) to obtain

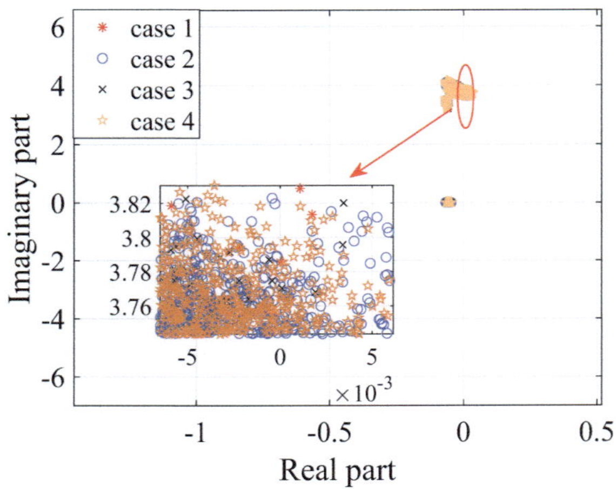

Fig. 6.10 The change trajectory of the spectral abscissa $\tilde{\sigma}(A)$ under different cases

the perturbation set Ω and the stable interval (LB, UB). The set Ω satisfies $\varphi \geq \tau$, where τ is 0.75. Therefore, perturbing the branches of the set Ω within the stable interval (LB, UB) also can guarantee the effectiveness of phyMTD.

To validate the above conclusion, we consider a more severe branch perturbation case under Ω. That is, we randomly select an eigenvalue, take the sign of its eigenvalue sensitivity as the perturbation direction, and perturb the branch reactances randomly within the stable interval (LB, UB). Based on the more severe cases, we repeat 10,000 experiments on the IEEE 39-bus and 68-bus test systems. Meanwhile, we plot the trajectories of the spectral abscissas. The results are shown in Fig. 6.11. We can see that none of the spectral abscissas is greater than 0, indicating that the small signal stability is retained.

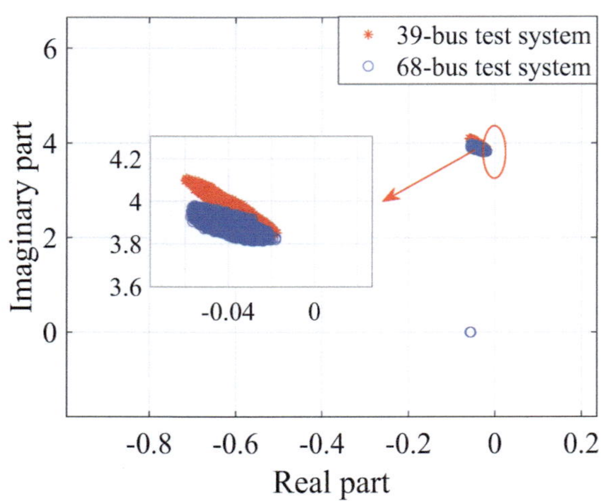

Fig. 6.11 The changing trajectory of the spectral abscissa $\tilde{\sigma}(A)$ under the stable interval (LB, UB)

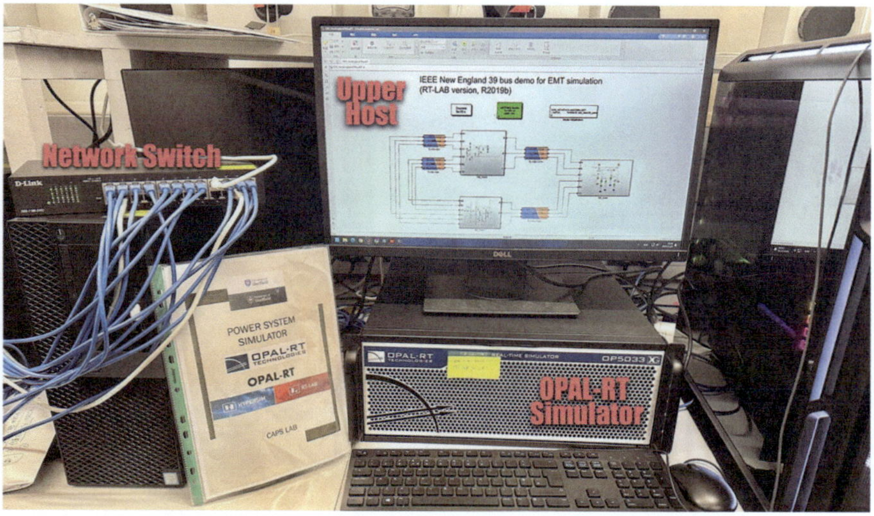

Fig. 6.12 The experimental platform of phyMTD with the OPAL-RT simulator

To make our findings more convincing, we use the OPAL-RT platform[2] to realize the real-world operation of phyMTD on the New England 39-bus power system. The benchmark network is located in New England, USA. It consists of 39 buses including 10 generator buses and 19 load buses. The OPAL-RT provides a high-fidelity simulation environment. It can accurately model the dynamic characteristics of the power system, including generators, transmission lines, and loads. The system's real-time simulation capability enables it to simulate the operation behaviors of the power system after phyMTD. The experimental platform is shown in Fig. 6.12. In particular, the high-fidelity OPAL-RT real-time simulator OP5033XG is used to emulate the electromagnetic transient dynamics of the New England 39-bus power system, where the upper host software RT-LAB is run in a high-performance computer equipped with 13-th Intel i9 CPU and 96GB RAM and exchanges model/monitor related data with the target machine via an Ethernet network switch. All generators are equipped with AVR and PSS. The AVR model is the DC1A exciter. The multi-band power system stabilizer (MB-PSS) is used as the stabilizer with the ARTEMIS library. We observe the rotor speed of the generator when all branches that were obtained by solving the optimization problem were perturbed, as shown in Fig. 6.13. The results show that the system is stable although the all branches are perturbed with phyMTD. Here, $\omega 1$-$\omega 10$ denote the rotor angle deviations of 10 generators, corresponding to bus 30-bus to 39-bus respectively. The details are introduced in Table 6.1.

To further demonstrate the above conclusion, we conduct simulations with the IEEE 68-bus power system [2]. Under Ω which is obtained by solving the optimization problem (6.22), we consider a most severe case where the sign of the

[2] https://www.opal-rt.com/.

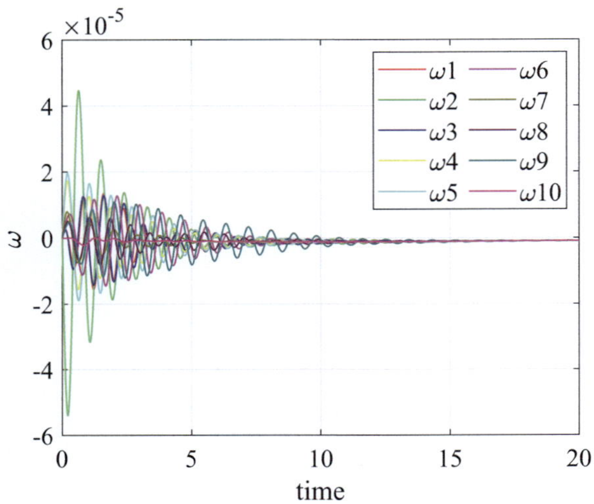

Fig. 6.13 The waveform of the deviation of the generator rotor speed when all branches that were obtained by solving the optimization problem were perturbed

Table 6.1 The rotor angle deviation and generator comparison table

Generator (G)	Bus	ω	Generator (G)	Bus	ω
1	39	$\omega 1$	6	35	$\omega 6$
2	31	$\omega 2$	7	36	$\omega 7$
3	32	$\omega 3$	8	37	$\omega 8$
4	33	$\omega 4$	9	38	$\omega 9$
5	34	$\omega 5$	10	30	$\omega 10$

eigenvalue sensitivity is taken as the perturbation direction. The branch reactance is perturbed with either the upper bound UB or the lower bound LB. We define the more and most severe cases as scenario 1 and scenario 2, respectively. The branch reactances are perturbed at $T = 20s$ and $T = 40s$ in these two scenario, respectively. Figure 6.14 shows the deviation of the generator's rotor speed after phyMTD. We can see that the deviation of the generator's rotor speed gradually approaches 0 after phyMTD, which indicates that the small signal stability is retained. Although scenario 2 is considered, the small signal stability is guaranteed. Therefore, we can realize an effective phyMTD without violating the system's small signal stability.

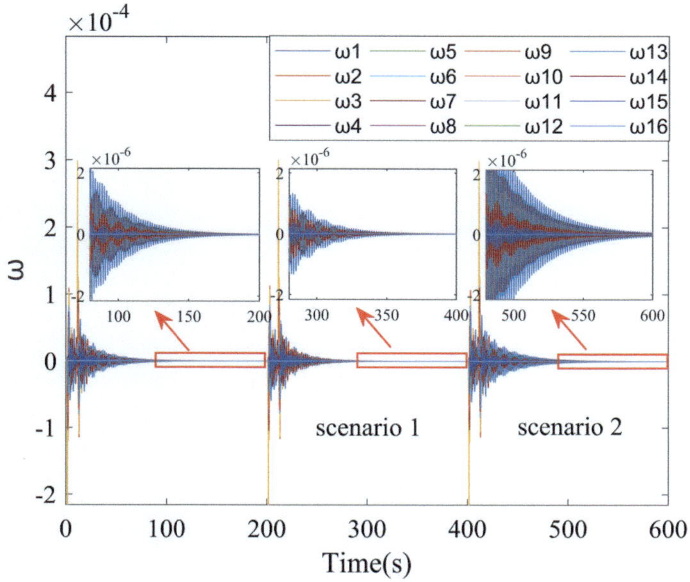

Fig. 6.14 The waveform of the deviation of the generator rotor speed

6.4.4 Improvement of Stability Margin

Here we validate that the margin of the small signal stability can be improved with the phyMTD. The phyMTD designed according to (6.34) is evaluated. The value of φ satisfies $\varphi \geq 0.75$. The D_1 and D_2 are two traditional phyMTDs with the branch reactances randomly perturbed within the perturbation limitation of $\pm 20\%$. We take the minimum damping ratio as the stability margin of the system. The simulation results are presented in Table 6.2. Without perturbation, the stability margin of the 39-bus and 68-bus test systems is 0.009848 and 0.010633, respectively. When the branches are perturbed by D_1 and D_2, the stability margin decreases to a certain extent. However, when the branch is perturbed with the phyMTD designed according to the optimization problem (6.34), the stability margin increases to 0.012241 and 0.020543, respectively. The results indicate that we can improve the stability margin by optimizing the construction of phyMTD.

Table 6.2 The damping ratio by solving the optimization (6.34) with respect to D_1 and D_2

	Without perturbation	D_1	D_2	Solving the optimization (6.34)
39-bus system	0.009848	0.007533	0.008835	0.012241
68-bus system	0.010633	0.008126	0.008589	0.020541

Table 6.3 The electromechanical modes with normalized participation factors of highest participating states

No.	Damping ratio (%)	State				Participation factor			
1	−5.16101	$\delta(13)$	$\omega(13)$	$\delta(12)$	$\omega(6)$	1	0.82527	0.14113	0.11821
2	−4.07915	$\delta(9)$	$\omega(9)$	$E_{qd}(9)$	$E_{fd}(9)$	1	0.99951	0.15049	0.06682
3	−1.02819	$\delta(13)$	$\omega(13)$	$\omega(15)$	$\omega(14)$	1	0.74438	0.54032	0.51151
4	0.25913	$\delta(2)$	$\omega(2)$	$\delta(3)$	$\omega(3)$	1	0.99178	0.91201	0.90350
5	0.62022	$\delta(14)$	$\omega(16)$	$\omega(14)$	$\delta(13)$	1	0.74648	0.49790	0.15389
6	2.53047	$\delta(10)$	$\omega(10)$	$\delta(1)$	$\omega(1)$	1	0.98840	0.63463	0.62417
7	3.15997	$\delta(12)$	$\omega(12)$	$\omega(13)$	$\delta(13)$	1	0.98504	0.19427	0.18088
8	3.27898	$\delta(10)$	$\omega(10)$	$\omega(1)$	$\delta(1)$	1	0.99356	0.68443	0.67837
9	3.32192	$\delta(15)$	$\omega(15)$	$\omega(14)$	$\delta(14)$	1	0.75530	0.30473	0.14864
10	3.63405	$\omega(2)$	$\delta(2)$	$\delta(3)$	$\omega(3)$	1	0.99793	0.81203	0.80999
11	3.76514	$\omega(5)$	$\delta(5)$	$\delta(6)$	$\omega(6)$	1	0.99874	0.81883	0.81749
12	3.90694	$\delta(11)$	$\omega(11)$	$\psi_{1d}(11)$	$\omega(10)$	1	0.99337	0.05647	0.03350
13	6.54849	$\delta(8)$	$\omega(8)$	$\omega(1)$	$\delta(1)$	1	0.99974	0.44705	0.44684
14	6.79901	$\delta(7)$	$\omega(7)$	$\omega(6)$	$\delta(6)$	1	0.99998	0.55681	0.55679
15	7.03255	$\delta(4)$	$\omega(4)$	$\omega(5)$	$\delta(5)$	1	0.99997	0.48340	0.48338

Moreover, we analyze if we can improve the stability margin when the system is in the electromechanical mode. The simulation results are given in Table 6.3. In this case, the damping ratio is less than 10% and the frequency is in the range 0.1–2 Hz. The simulation results are shown in Fig. 6.15. We can see that the frequency is maintained in the range with M_1 and M_2, while the stability margin is increased from $\xi_{min} = 0.0106$ to $\xi_{min} = 0.0205$ with M_2, which indicates that the stability margin is improved by solving the optimization (6.34).

6.5 Summary

In this chapter, we deeply analyzed the impact of phyMTD on the small signal stability. To begin with, we derived the relationship between the small signal stability and the perturbation of branch reactance using the theory of eigenvalue sensitivity. Then, we proposed a new indicator to measure the effectiveness of phyMTD. Further, we compute the stable interval of the reactance perturbation by formulating an optimization problem with the constraint of retaining the small signal stability. Besides, we improved the stability margin by increasing the damping ratio. Finally, we performed extensive simulations to evaluate the performance of the designed phyMTD.

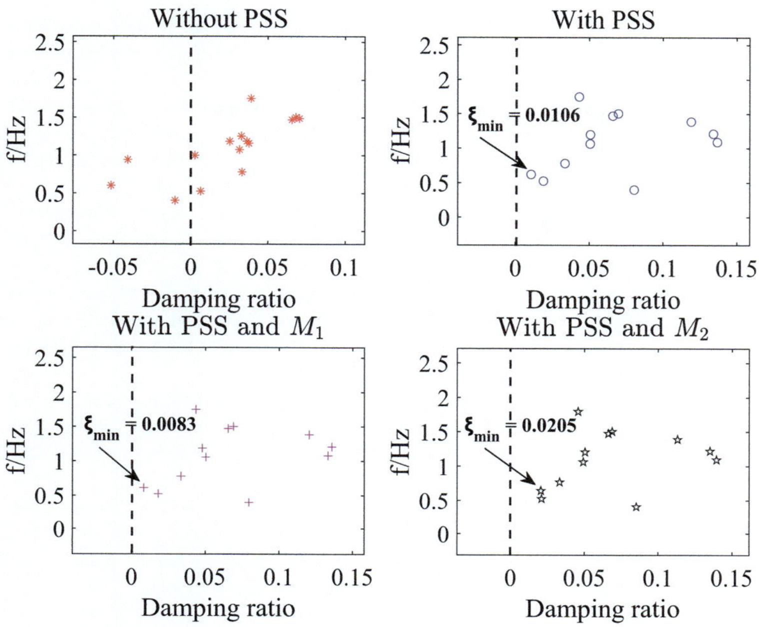

Fig. 6.15 The relationship between damping ratio and frequency under different conditions. Where M_1 denotes a random perturbation in the stability interval (LB, UB) obtained by solving the optimization problem (6.22), while M_2 denotes the perturbation using Ω^* and ΔX^* obtained by solving the optimization problem (6.34)

References

1. M.S. Abdalzaher, M.M. Fouda, M.I. Ibrahem, Data privacy preservation and security in smart metering systems. Energies **15**, 19 (2022)
2. K.S. Abhinav, C.P. Bikash, IEEE PES Task Force on Benchmark Systems for Stability Controls: Report on the 68-Bus, 16-Machine, 5-Area System. IEEE PES (2013)
3. Israeli Hackers Cause Major Disruptions in Iranian Electricity Grid. https://www.time.news/israeli-hackers-cause-major-disruptions-in-iranian-electricity-grid/
4. Z. Zhang, B. Wang, M. Liu, Y. Qin, J. Wang, Y. Tian, J. Ma, Limitation of reactance perturbation strategy against false data injection attacks on IoT-based smart grid. IEEE Internet Things J. **11**(7), 11619–11631 (2024)
5. S. Lakshminarayana, E.V. Belmega, H.V. Poor, Moving-target defense against cyber-physical attacks in power grids via Game theory. IEEE Trans. Smart Grid **12**(6), 5244–5257 (2021)
6. B. Li, G. Xiao, R. Lu, R. Deng, H. Bao, On feasibility and limitations of detecting false data injection attacks on power system state estimation using D-FACTS devices. IEEE Trans. Industr. Inf. **16**(2), 854–864 (2020)
7. Z. Liu, Y. Li, Q. Wang, J. Li, Moving target defense of FDIAs for battery energy storage systems in smart distribution networks. J. Energy Stor. (2023)
8. B. Liu, H. Wu, Optimal planning and operation of hidden moving target defense for maximal detection effectiveness. IEEE Trans. Smart Grid **12**(5), 4447–4459 (2021)
9. S.G. Marín, W.G. Vanegas, C.E.M. Sánchez, MPNG: A MATPOWER-based tool for optimal power and natural gas flow analyses. IEEE Trans. Power Syst. 1–9 (2022)

10. K.L. Morrow, E. Heine, K.M. Rogers, R.B. Bobba, T.J. Overbye, Topology perturbation for detecting malicious data injection, in *45th Hawaii International Conference on System Sciences*, Maui, HI, USA (2012), pp. 2104–2113
11. D. Moutevelis, G. Tzounas, J.R. Perez, F. Milano, Modal propagation analysis with participation factors of complex frequency variables. Electric Power Syst. Res. (2024)
12. D.T. Peng, J. Dong, J. Yang, Q. Peng, Dynamical failures driven by false load injection attacks against smart grid. IEEE Trans. Inf. Forensics Secur. **17**, 2213–2226 (2022)
13. Sandworm Disrupts Power in Ukraine Using a Novel Attack Against Operational Technology. https://cloud.google.com/blog/topics/threat-intelligence/sandworm-disrupts-power-ukraine-operational-technology/
14. P.W. Sauer, M.A. Pai, J.H. Chow, Small-signal stability, in *Power System Dynamics and Stability: With Synchrophasor Measurement and Power System Toolbox* (IEEE, 2017), pp. 183–231
15. S. Sivasubramani, EE549 - power system dynamics and control small signal stability. https://www.iitp.ac.in/~siva/2022/ee549/Small_Signal_Stability.pdf
16. J. Tian, R. Tan, X. Guan, T. Liu, Enhanced hidden moving target defense in smart grids. IEEE Trans. Smart Grid **10**(2), 2208–2223 (2019)
17. Z. Zhang, R. Deng, Impact analysis of MTD on the frequency stability in smart grid. IEEE/CAA J. Autom. Sinica **10**(1), 275–277 (2023)
18. H. Zhang, B. Liu, X. Liu, A. Pahwa, H. Wu, Voltage stability constrained moving target defense against net load redistribution attacks. IEEE Trans. Smart Grid **13**(5), 3748–3759 (2022)
19. Z. Zhang, R. Deng, D.K. Yau, Vulnerability of the load frequency control against the network parameter attack. IEEE Trans. Autom. Control **15**(1), 921–933 (2023)
20. Z. Zhang, R. Deng, D.K.Y. Yau, P. Cheng, M.-Y. Chow, Security enhancement of power system state estimation with an effective and low-cost moving target defense. IEEE Trans. Syst. Man Cybernet.: Syst. **53**(5), 3066–3081 (2023)

Chapter 7
Summary and Future Directions

7.1 Summary

This monograph provides a comprehensive analysis of MTD in the smart grid, focusing on its effectiveness, hiddenness, practical implementation, and stability impacts. Through theoretical analysis and extensive simulations on IEEE standard test systems, we have demonstrated that MTD can effectively protect power systems against FDI attacks while maintaining system stability. The key findings and contributions are summarized below.

7.1.1 Effectiveness in Detecting False Data Injection Attacks

In Chap. 3, we conducted a comprehensive analysis of MTD's effectiveness against stealthy FDI attacks under the DC power flow model. We established key theoretical results regarding MTD completeness, showing that an MTD scheme can defeat all FDI attacks based on outdated system information if: (1) the number of branches l is at least twice the number of system states n (i.e., $l \geq 2n$, where $n + 1$ is the number of buses), and (2) susceptances are perturbed on more than n branches that cover all buses. We proved that complete MTD is impossible for systems with single-branch connected buses. Importantly, we demonstrated that the magnitude of susceptance perturbations has minimal impact on the dimension of the post-MTD stealthy attack space. These findings led to practical guidelines for MTD implementation that minimize attack opportunities while optimizing bus coverage and operational costs. Our theoretical results were validated through extensive simulations on IEEE standard test systems.

R. Deng et al., *Moving Target Defense in the Smart Grid*, Wireless Networks,
https://doi.org/10.1007/978-3-031-92378-4_7

7.1.2 Hiddenness from Attackers

In Chap. 4, we explored the conditions necessary for achieving undetectable MTD. We classified attackers into two categories based on their knowledge level: partially knowledgeable and fully knowledgeable attackers. Using an MTD-confirming detector similar to BDD, we proved that MTD can only potentially remain hidden from partially knowledgeable attackers. For fully knowledgeable attackers who possess complete information about branch parameters and measurements, we demonstrated that any effective MTD scheme becomes detectable. To address this fundamental limitation, we developed a novel MPS+MTD co-design strategy. By protecting a specific set of measurements, this approach achieves both hiddenness and effectiveness regardless of power flow variations, attacker knowledge, or branch perturbations. Extensive simulations on IEEE test systems validated these theoretical findings.

7.1.3 Explicit Design in AC Power Systems

In Chap. 5, we developed EXR-MTD, a novel approach for explicit MTD analysis in AC power systems using measurement residuals. Our method employs sensitivity analysis around the optimum point to derive explicit approximations of residuals, enabling joint optimization of MTD effectiveness and hiddenness. Simulation results demonstrated that EXR-MTD achieves superior hiddenness compared to existing strategies while maintaining comparable or better effectiveness. The approach showed practical viability through two key findings: (1) minimal impact on OPF generation costs due to limited power flow fluctuations, and (2) acceptable computation times even for large-scale systems with over 2000 buses. We also identified unique characteristics of power distribution systems, where small X/R ratios provide natural hiddenness, suggesting the need for additional metrics beyond effectiveness and hiddenness for MTD design in these contexts.

7.1.4 Small Signal Stability Analysis

In Chap. 6, we conducted an in-depth investigation of physical MTD's impact on small signal stability. Using eigenvalue sensitivity theory, we established the mathematical relationship between small signal stability and branch reactance perturbations. We introduced a novel metric for quantifying physical MTD effectiveness and formulated an optimization problem to determine stable intervals for reactance perturbations while maintaining system stability. Our analysis revealed opportunities to enhance stability margins through damping ratio optimization. Through comprehensive simulations, we validated the performance characteristics of our stability-aware physical MTD design approach.

7.2 Future Directions

While significant progress has been made in developing MTD schemes for smart grid security, several important research directions remain to be explored. These include integrating cyber and physical MTD capabilities through co-design approaches, establishing formal standards and guidelines for practical MTD implementation, and developing intelligent MTD agents that can adapt to dynamic adversarial environments. Below we discuss these directions in detail.

7.2.1 Cyber-Physical Moving Target Defense Co-design

The modern smart grid presents unique opportunities for integrating cyber and physical MTD capabilities. On the cyber side, software-defined networking controllers enable flexible reconfiguration of communication paths and protocols. On the physical side, D-FACTS devices and digital controllers allow dynamic adjustment of power system parameters and operating points. However, these cyber and physical MTD mechanisms have largely been studied in isolation. A promising direction is to develop comprehensive co-design frameworks that coordinate cyber and physical MTD actions. This requires modeling the interdependencies between cyber and physical layers, analyzing how MTD actions in one layer impact the other, and designing joint optimization approaches. Key challenges include maintaining system stability and performance while maximizing security benefits through coordinated cyber-physical MTD. Success in this direction could significantly enhance grid resilience against sophisticated cyber-physical attacks.

7.2.2 Moving Target Defense Operation Standard

For MTD to transition from academic research to practical deployment, formal standards and guidelines are essential. These standards must address multiple aspects of MTD operation while ensuring grid reliability and stability. First, they should specify which devices and components are suitable for MTD perturbations, considering factors like device capabilities, operational constraints, and potential impacts on system performance. Second, standards must define clear conditions and criteria for when MTD actions can be triggered, including thresholds for system states, detection of suspicious activities, and coordination with other security mechanisms. Third, they should establish requirements for perturbation frequencies, magnitudes, and patterns that balance security benefits with operational stability. Developing these standards requires close collaboration between academia, industry, and regulatory bodies to ensure they are both theoretically sound and practically implementable. The standards should also be flexible enough to accommodate technological advances while maintaining backward compatibility with existing infrastructure.

7.2.3 Intelligent Moving Target Defense Agent

The dynamic nature of cyber threats demands MTD schemes that can adapt to evolving attack strategies and changing system conditions. Static MTD approaches with fixed patterns or predetermined rules are vulnerable to detection and circumvention by sophisticated adversaries. Developing intelligent MTD agents powered by advanced machine learning techniques, particularly reinforcement learning, represents a promising solution. These agents would continuously learn from system observations, attack patterns, and defense outcomes to optimize MTD strategies in real-time. Key research challenges include designing appropriate state and action spaces, developing reward functions that balance security and performance objectives, and ensuring learning convergence under partial observability. The agents should also be capable of transfer learning to quickly adapt to new scenarios and attack variants. Additionally, research is needed on multi-agent coordination when multiple intelligent MTD agents operate in different parts of the grid.

7.2.4 Moving Target Defense Incentive Mechanism

The successful implementation of MTD in smart grid systems requires active participation from multiple stakeholders, including DER owners, customers, and system operators. A well-designed incentive mechanism is essential to encourage stakeholder participation while ensuring fair compensation for their contributions to grid security. This presents several key research challenges. First, appropriate pricing models need to be developed that quantify the security value provided by different MTD actions and translate this into monetary incentives. Second, market mechanisms must be designed to efficiently allocate MTD resources and rewards among competing participants. Third, the incentive structure should promote long-term commitment to MTD participation while remaining flexible enough to adapt to changing grid conditions and security needs. A promising direction is to establish MTD as a formal grid-support service, similar to existing ancillary services like frequency regulation and voltage support. This would enable integration with existing market frameworks while providing clear value propositions to stakeholders. Research is needed to determine optimal reward structures that balance security benefits against implementation costs. Additionally, as MTD competes with other grid services for limited resources, new optimization approaches are required to maximize overall system welfare through coordinated service allocation. Success in this direction could significantly accelerate MTD adoption by creating sustainable economic incentives for all participants.

MIX
Papier aus verantwortungsvollen Quellen
Paper from responsible sources
FSC® C105338

If you have any concerns about our products, you can contact us on
ProductSafety@springernature.com

In case Publisher is established outside the EU, the EU authorized representative is:
Springer Nature Customer Service Center GmbH
Europaplatz 3, 69115 Heidelberg, Germany

Printed by Libri Plureos GmbH
in Hamburg, Germany